William J. Tennant, John Henry Kinealy

The Slide Valve

Simply Explained

William J. Tennant, John Henry Kinealy

The Slide Valve
Simply Explained

ISBN/EAN: 9783337379520

Printed in Europe, USA, Canada, Australia, Japan

Cover: Foto ©berggeist007 / pixelio.de

More available books at **www.hansebooks.com**

THE

SLIDE VALVE,

SIMPLY EXPLAINED.

By W. J. TENNANT, Asso. M.I.M.E

REVISED AND MUCH ENLARGED

By J. H. KINEALY, D. E.,

Professor of Mechanical Engineering in Washington University;
M. Am. Soc. Mech. Engrs. Author of "Steam Engines and
Boilers;" "Low Pressure Steam Heating Charts."

FULLY ILLUSTRATED WITH ORIGINAL DRAWINGS AND DIAGRAMS.

SPON & CHAMBERLAIN,
12 CORTLANDT STREET, NEW YORK.
1899.

PREFACE.

THIS work is based upon notes and diagrams which were prepared by the writer with no more ambitious object, originally, than to help his railway students toward the obtainment of clear *general* notions upon the important subject of the slide valve.

For practical demonstration of the functions and operation of the slide valve, a large sectional model engine, with provision for variation of the operation of its parts, is very valuable; but it seems to the writer to be desirable that each student should have in his own possession a model to cogitate upon, and that his model should give him, graphically, results which should be not merely qualitative, but also, in some degree, *quantitative,* to enable him to institute comparisons between the actions of different valves operated under varying conditions. With this end in view, the writer conceived the idea of using on a base-board a rotary disc to represent a crank-shaft, together with the idea of obtaining concentric circular diagrams of results (see fig. 21, for instance) by using a crank-arm marked on the disc as an index-finger, and recording on the base-board the beginnings and ends of the arcs swept through by the crank in the various distribution-periods. Following upon this, it was thought that the eccentric and its rod could be eliminated by bringing the disc under the valve and fixing upon the valve a rigid slotted arm to extend down across the face of the disc and engage a pin on the latter, so that rotation of the disc would cause reciprocation of the slotted arm and the valve. This promised, however, to be unmechanical and expensive, and was discarded in favor of the model described in Chapter I., which comprises the simple method of correlating (as described in Chapter II.) the movements of valve and crank-shaft by scales adjacent to each, so that the reading given by the valve on the valve-scale at any position of the valve would show what position on the eccentric-scale the eccentric ought correspondingly to occupy. The fact that this involved a step-by-step adjustment of the valve and shaft alternately was found in practice to be advantageous rather than the reverse.

As to the diagrams and sketches, the writer has sought, in some of them, to clear up graphically some of the incidental minor difficulties related to the subject; see, for instance, figs. 12 and 13, illustrating " Order of Cranks," fig. 22 on " Width of Port "; also compare figs. 23, 24, and 25, " Double-ported Valve," and see fig. 37, which last is in effect a graphical summary of results from nearly all the different examples which precede it.

Fig. 36, which concerns the reversal of an engine in motion under steam, will probably have a special interest to students of the locomotive.

W. J. T.

CONTENTS.

INTRODUCTION TO AMERICAN EDITION.

There is no more important part of the steam-engine than the valve—the part which determines when and for how long the steam shall be admitted to the cylinder, how long it shall stay there and how long it shall be allowed for leaving. And the principles which should govern the construction of the valve of an engine in order that the steam shall be properly admitted and released are the same for American, English, French and German engines. The principles of their construction and action are the same, but the details vary. The action of every steam-engine depends upon its valve, and hence it is important that every engineer of whatever country should be thoroughly conversant with the principles of the construction of valves of different types; should know the meaning of " lap," " lead," " advance," etc.; and should know how a change of one of these will affect the others and the admission and release of the steam. One of the best ways to learn all about valves and valve motions is to study them with the aid of sectional models, made so that the inside as well as the outside of the parts may be seen, but such models are difficult to obtain except at considerable cost. Diagrams, such as are so freely used in this work, rank next to models in utility and as a means of enabling one to thoroughly understand the workings of the different parts of machines. The use of diagrams requires no knowledge of mathematics, and enables explanations to be presented in a clear, concise manner. From an educational point of view, diagrams possess an advantage which models do not, as they make one think more. Their use gets one in the habit of forming mental pictures of the parts of the machines discussed, and thus enables one to more readily and quickly think out the effect of a change in one part on other parts. And this ability to picture in the mind the various relations of the parts to one another is absolutely necessary in order that these relations may be thoroughly understood. The man who is accustomed to work out his problems from diagrams and drawings, reasons from cause to effect and from effect to cause; while the man who must have a model to work with, works on the " cut-and-try " method.

This little book is confined strictly to an explanation of the principles which underlie the action of the different types of slide valves. The plain, simple D-valve, as it is called in this country, without lap or lead, is first taken up and discussed; and gradually lap and lead are introduced, and the effect of each upon the admission, the cut-off, the release, and the compression is fully worked out and shown. Then other types of slide valves are taken up and discussed. The advance is made step by step, from the most simple form of slide valve to the more complicated forms used for attaining certain results not so easily attained by the use of the simple forms.

The form of multi-ported valve shown in fig. 25 is so much like the Giddings valve used on the Russell and some other single-valve engines, that any one reading the explanation given here will have no trouble in understanding the Giddings valve (shown in fig. 26).

The multiple admission valve is used to such an extent in America that it is but proper that a few words should be said of it, and that the most common form of multiple admission valves should be discussed. And hence the Straight Line and the Woodbury valves are shown in figs. 27 and 28.

There are so many engines in this country in the front rank of automatic, high-speed engines which use piston-valves, that a treatise on slide valves would be incomplete without some mention of this type of slide valve. The form of piston-valve shown in fig. 30, in which the steam is admitted at the middle and exhausted at the ends, is well known to all engineers who have used an Ide or an Ideal engine. The valve used on these engines is shown in fig. 31.

Many American engineers prefer engines on which the cut-off may be changed without in any way affecting the lead, release, or compression, and, therefore, the description of the Meyer valve will be read with interest. Many will recognize at once in the valve used on the Watertown engine the Meyer valve in almost its original form, and in the valve of the Buckeye engine (shown in section in fig. 39), the Meyer valve in a modified form.

J. H. KINEALY, D.E.,

Prof. of Mech. Eng., Washington Univ.,
St. Louis, Mo.

THE SLIDE VALVE.

CHAPTER I.

THE SIMPLE SLIDE.

THE slide valve has always been regarded by the writer as the *pons asinorum* of the student of the steam-engine. His own early attempts at crossing that bridge were greatly facilitated by the simple little device to be presently submitted to the reader; with the aid of this expedient the student may easily obtain clear notions upon all the functions and operations of the slide valve, at a nominal cost, and with but a small expenditure of time and trouble.

It is necessary that at the outset the nature of the construction above alluded to should be clearly appreciated, for the reason that many of the explanations given in the succeeding pages are referred thereto.

In passing, it may be remarked that much of the difficulty which the ordinary student experiences in the study of valves and valve gear arises because he approaches the subject

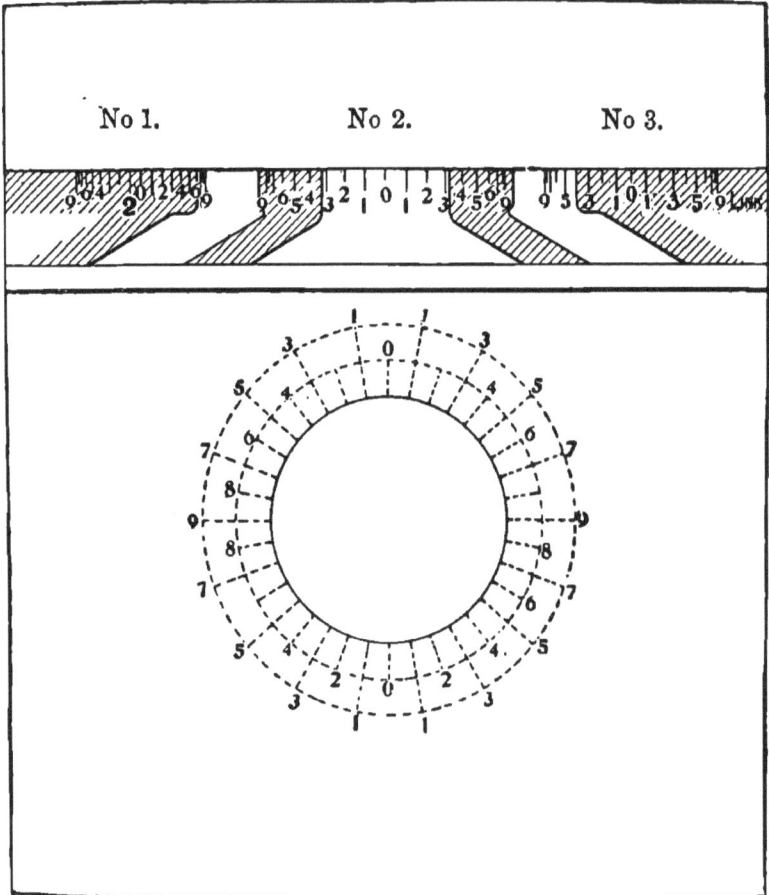

FIG. 1.—Explanatory Diagram.

with a false impression that it is of necessity a difficult one; this idea is generated on the one hand in the atmosphere of mystery which in the shop is made to surround the matter, and on the other hand by the somewhat formidable aspect of most of the geometrical treatises on the subject. If the beginner can bring himself to believe that the slide valve is nothing more complex than a sliding shutter with a cavity in its face, travelling backwards and forwards over three ports, the outermost of which are alternately opened to steam and placed in communication with the central exhaust-port, he will have made a satisfactory commencement.

Take a piece of stout white cardboard and upon it set out an enlarged copy of fig. 1.*

The upper portion of this diagram represents a section taken at right angles to the port-face through a set of steam and exhaust-ports as ordinarily arranged; the graduated ring beneath the section will receive consideration later.

Postponing for the moment the further investigation of fig. 1, reference should here be made to figs. 2, 3, and 4.

Fig. 2 shows, in perspective, a cylinder and valve-chest with parts removed to make clear the manner in which the two steam-ports

* See also fig. 41.

(S_1S_2) at the upper ends of the two outer passages (P_1P_2) afford a communication between
the valve-chest (VC) and the opposite ends
(C_1C_2) of the cylinder (C), and to show the
passage from the central exhaust-port (EP) to

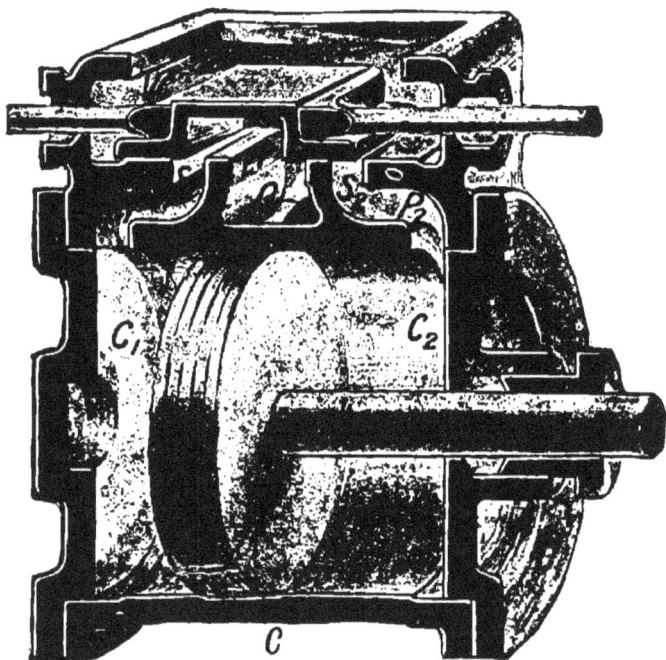

FIG. 2.—Perspective Section of Steam-Engine Cylinder.

the outlet (O) at which the exhaust steam is
discharged. The relationship between this view
and the upper part of fig. 1 will be obvious
upon a comparison of the two.

A typical slide valve is shown in section, in

its place above the ports in fig. 2, and separately
in figs. 3 and 4. From these latter, the large
ratio of the width to the length of the valve is

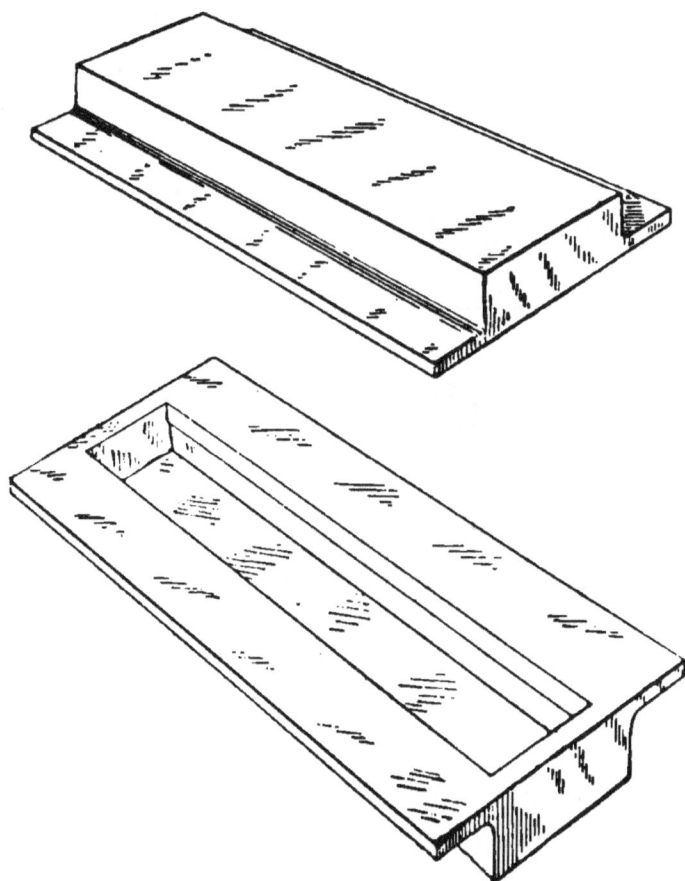

FIGS. 3 and 4.—Separate Views of the Slide Valve.

apparent, and it will be seen from fig. 2 that
the ports are correspondingly made wide and
short. This arrangement enables a small linear

movement of the valve to effect the uncovering of a large area of port, and to secure free
passage for the steam with but a small amount
of valve-travel, and so to reduce to a minimum
the work unavoidably wasted in overcoming
friction between the port-face and the valve
heavily loaded by the pressure of steam upon
its back.

Let us return now to fig. 1. Draw, near the
edge of a strip of moderately stiff paper, the ·

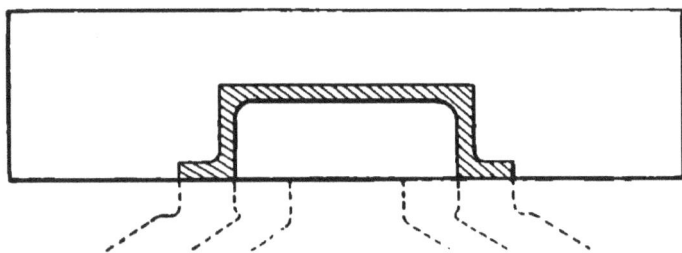

FIG. 5.—Section of Slide Valve.

section of the elementary form of slide valve
illustrated by fig. 5, taking its dimensions from
the ports of fig. 1, so that the two views agree
in the manner indicated. By sliding this diagram backwards and forwards across the ports
of fig. 1, the following explanation of the action
of this, the simplest form of slide valve, will
be readily appreciated.

This valve has first to open one port to

steam, and when the steam thus admitted has forced the piston towards the opposite end of the cylinder, to put the same port in communication with the exhaust, while opening the other port to the steam which effects the return movement of the piston. The duration of the admission of steam is the same as that of the exhaust, the admission, owing to the dimensions of the valve in relation to the ports, neces-

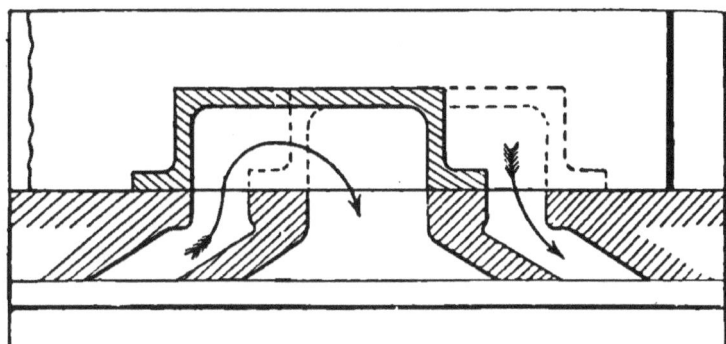

FIG. 6.—Section showing Position of Slide Valve in relation to Steam and Exhaust-Ports.

sarily taking place on one side of the piston simultaneously with the occurrence of exhaust from the other side (see fig. 6). The travel of this rudimentary valve equals twice the amount to which the port is opened to steam. It happens to open each port fully in this case (although this is rarely the case with modern slide valves), so that the "travel" equals twice the width of steam-port : *twice*, because (starting with the valve in its central position, as shown in fig. 5) it has to move to the right, say, by an amount equal to the *width* of *the left-hand* port, in order to open that port fully to steam, and then it moves back again and travels to the other side of the central position by an equal amount, in order to open the equally dimensioned right-hand port to steam. In fig. 6 the valve is shown at the two extremes of its travel, in full lines at one end, and in dotted lines at the other end.

CHAPTER II.

THE necessary travel of the valve is given to
it by means of an eccentric, which is keyed
upon or formed as part of the crank-shaft. The
eccentric is, in effect, a crank, whose throw
equals the amount of eccentricity of the sheave.
This may not be obvious; let us investigate a
little. Suppose that we have a big crank-shaft,
and want to put a little crank in the middle of
it (for this is what happens in the case of a
steam-engine, the travel of the valve being such
that a crank of the usual ⊓ form
would be disproportionate to the shaft of which
it formed part—like fig. 7, perhaps). Now,
that such a shaft would be extremely weak at
the pin goes without saying. Imagine, that
to lessen the weakness the crank-pin of this
little crank is made of greater diameter, as in
fig. 8, or even more so, as in fig. 9; still we get
the same result, which is that the *throw* of the

crank remains the same, and equals the distance of the centre of the crank-pin from the centre of the crank-shaft—*i.e.,* the amount that this crank-pin is "*ex-centric,*" out of centre,

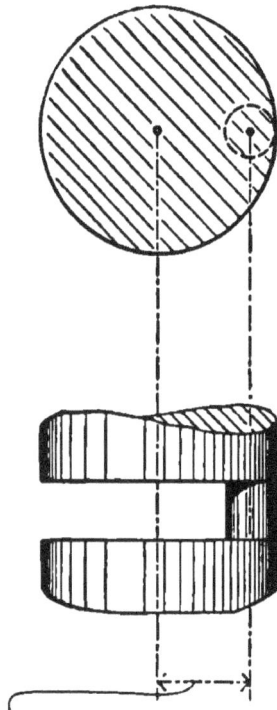

FIG. 7.—Half-travel of Valve.

and *the eccentric sheave being simply an exaggerated crank-pin* we get back to our original statement that it is virtually a crank whose throw equals the half-travel of the valve, which in turn equals the amount of eccentricity of the sheave.

Being fixed to the crank-shaft and moving with it, always in a fixed relationship to the

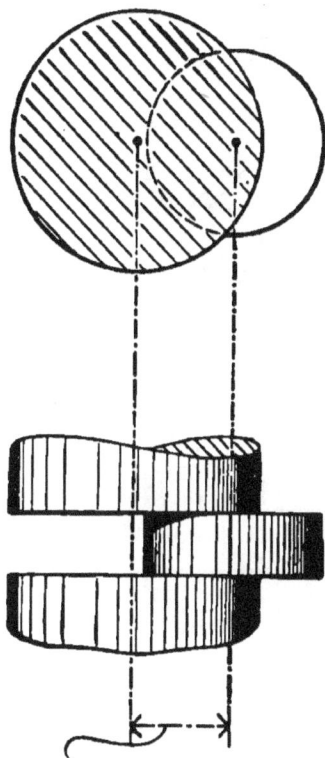

FIG. 8.—Half-travel of Valve.

main crank, the eccentric operates the valve in the necessary accordance with the movement of the crank and piston, as will presently be explained.

Returning now to fig. 5, it must be understood that a valve of the type exemplified

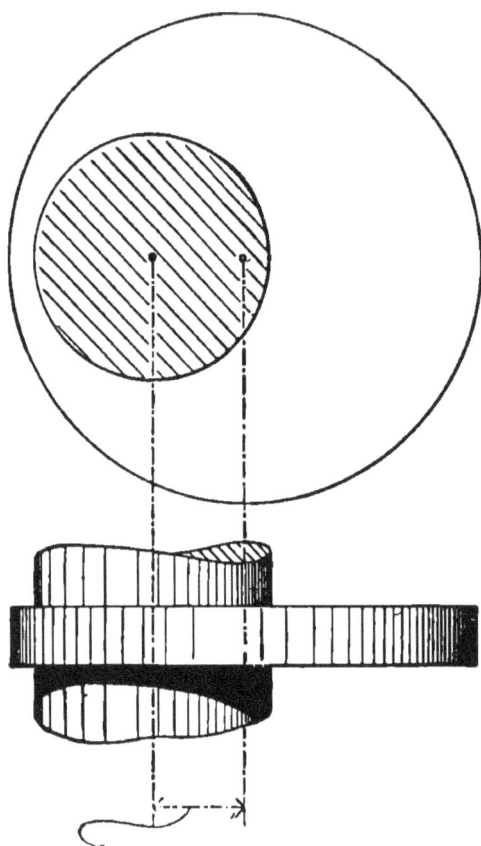

FIG. 9 — Half-travel of Valve.

therein must be in the *middle* of its travel
whenever the piston is at *either end* of its
stroke. The reason for this must be clearly

appreciated for the sake of what follows here-
after; it may be arrived at very easily with the
aid of fig. 1. Put a disc of card in the circle
provided in that figure, and let it turn about
a drawing-pin stuck through its centre; this
disc is the equivalent of a crank-shaft. A
crank-arm (C) may be permanently marked
upon it in ink, as in fig. 10; another arm, which
may be marked upon it in pencil, will serve
to represent a " valve-crank," the equivalent
of the eccentric by which the valve is to be
operated.

Actual connection between that valve-crank
and the valve being dispensed with, some
means for making the movement of the valve
correspond accurately with that of its crank
must be provided. This provision is made by
the numbered scales * in fig. 1; the circular
scale within which the card disc revolves is
numbered in correspondence with scales on the
ports, and to these latter an arrow on the strip

* The horizontal scales on the port-faces are obtained by
projecting up equidistant circumferential marks from various
circles having different valve-cranks for their radii. The scale
(No. 1) is for the elementary form of valve, without lap; it serves
also for link-motion in mid-gear ; No. 2 is for a valve having
outside lap ; and No. 3 is for link-motion "linked up." The
student need not, unless he pleases, troub e himself to investigate the construction of the diagram, which is drawn to scale,
so that if copied accurately it will serve to give him the dem
onstrations desired.

of paper which carries the valve is adjusted
(see also fig. 10). To find the position in
which to mark the arrow on the valve, put the
latter in the centre of its travel and mark the

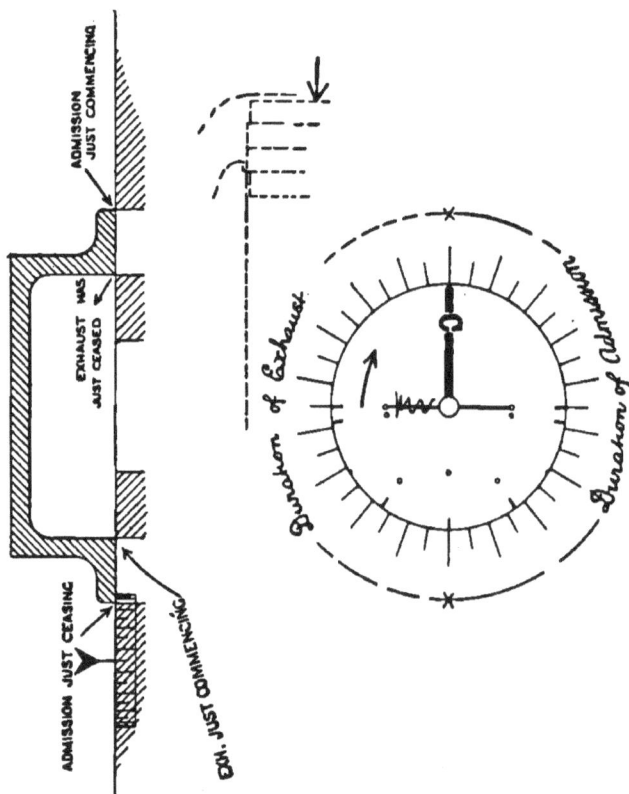

arrow-head just above the " O " on the No. 1
scale. When the disc is turned, the end of
the eccentric-arm or valve-crank travels with
it around the circular scale. When the eccen-

tric-arm is moved step by step around the circle to the various numbered graduations, the valve must be moved step by step to bring its arrow-head to the corresponding numbers on the horizontal scale on the ports. Thus the eccentric and valve, although not moved simultaneously, as they would be if a rod connected them, are kept in accurate relationship whilst moved independently. If the reader will now read the preface, which he has probably skipped, he will the more readily appreciate the principle of this arrangement.

Further, taking the main crank-arm (C) as an index-finger, the position of its end at the points of cut-off, release, etc., with different valves and eccentrics, may be marked in pencil on the circular scale, or better, on a piece of tracing-paper interposed between the disc and that scale; by this means the student will obtain circular diagrams enabling him to compare the results due to different settings and proportions of gear, and he will also discover in what manner those results are affected by individual elements of the mechanism in any given example.

CHAPTER III.

ASSUME a horizontal engine, with the main crank moved into a horizontal position so that the piston would be at, say, the left-hand end of its stroke. Assume also that the engine is at rest, and put the valve in the centre of its travel (as in fig. 5), when its faces exactly cover the steam-ports and its cavity covers the bridges and the exhaust-port; the first movement of the valve is wanted to be to the *right,* in order to admit steam to the left-hand end of the cylinder to move the piston also to the right, no matter whether the crank-shaft is to have negative or positive rotation. Now, a valve of this pattern evidently will not admit steam to the cylinder to *start* the movement of the piston (supposing that the engine has not yet been started), which will therefore have to be helped by some external agency; admitting this as a point which shall be touched again presently, place the crank next, so that the piston would be at the opposite end of its stroke; the valve, which had to move to the right as explained, must have come back into the position whence it started, and be in readiness to move equally to the *left,* afterwards returning once more to the central position by the time that the piston gets back to the left-hand end of the cylinder. Hence we see that, whenever the piston is at either end of its stroke, this valve must be in the middle of its travel.

If this be granted, we may proceed further: —Having the piston at the right-hand end of its stroke, for instance, draw a line on the disc at right angles with the crank (see fig. 10); this will represent two " eccentric-arms " of the necessary throw if each arm equals in length the half-travel of the valve; either of them at present suits the position of the valve, which is in the middle of its travel. Choose, now, the direction in which your crank-shaft shall rotate; let us suppose, for example, that it shall have " right-handed " rotation, and place an arrow on the disc to indicate the direction of rotation. Under these conditions the upper eccentric will *follow* the crank round, operating the valve to close the right-hand port and open the left-hand one, admitting steam to stop the engine. Cross this upper arm out, then, for manifestly we have to use the lower eccentric, which, going ahead of the crank, will cause the valve to open the right-hand port when required. But now conversely, suppose that the engine is to be run in the opposite direction; the upper eccentric (represented by the crossed-out line) now goes ahead of the crank, and gives the valve motion in the right direction, consequently the lower one must now go out of use, for under the altered conditions of working it tends to move the valve in the wrong direction. The deduction from this reasoning is, that in whichever direction an engine runs, the eccentric used for the time being must be set in advance of the crank, and that advance must be *at least 90°*.

If it were less than 90° the wrong port would be opened—the port at the opposite end of the cylinder to that at which the piston might happen to be. This effect may be shown by purposely giving the eccentric less advance.

The amount which the angle between the eccentric and the crank exceeds 90° is the " angle of advance "; and the distance which the valve is moved from mid-position when the piston is at the end of the stroke is the " linear advance."

Because of the necessity for different settings of the eccentric for different directions of rotation of the crank, engines required to be reversible usually have either:

(1) Means for shifting the position of a single eccentric upon the shaft, that it may always be placed so as to lead the crank. Or

(2) Two eccentrics fixed upon the crankshaft so that one of them is always ahead of the crank in whichever direction it rotates, mechanism being arranged in connection with the eccentrics, so that the valve can be driven by the leading eccentric, or operated by the conjoint action of both eccentrics. Or

(3) Radial or other special valve gears.

The simple form of slide valve shown in fig. 10 does not permit of the economical use of steam, inasmuch as it allows steam at full pressure to follow the piston for the whole of the stroke, and does not admit of the use of its expansive properties, for the simple reason that, at the instant the admission of steam ceases, the exhaust of the same body of steam must immediately commence, as may be clearly seen from the circular diagram.

CHAPTER IV.

It is desirable at this point to clear the ground by touching for a moment upon one or two elementary matters which, if not now explained, might cause confusion subsequently.

Dead Centres.—A crank is said to be " on the centre " or " on the dead centre " when the connecting-rod and crank are in line, and this occurs twice in every revolution, as shown in fig. 11.

Right-hand Crank to Lead.—The phrases " right-hand crank to lead," or " left-hand crank to lead," are sometimes used, and are confusing to beginners; the " leading " crank is that one which leads *when the engine is running ahead.* In most two-crank engines one of the cranks will be a quarter of a revolution (or *less* than *half* a revolution, where the cranks are not set at 90 degs.) in advance of its neighbour, and will, therefore, *lead* it in the direction in which it should go. Of course,

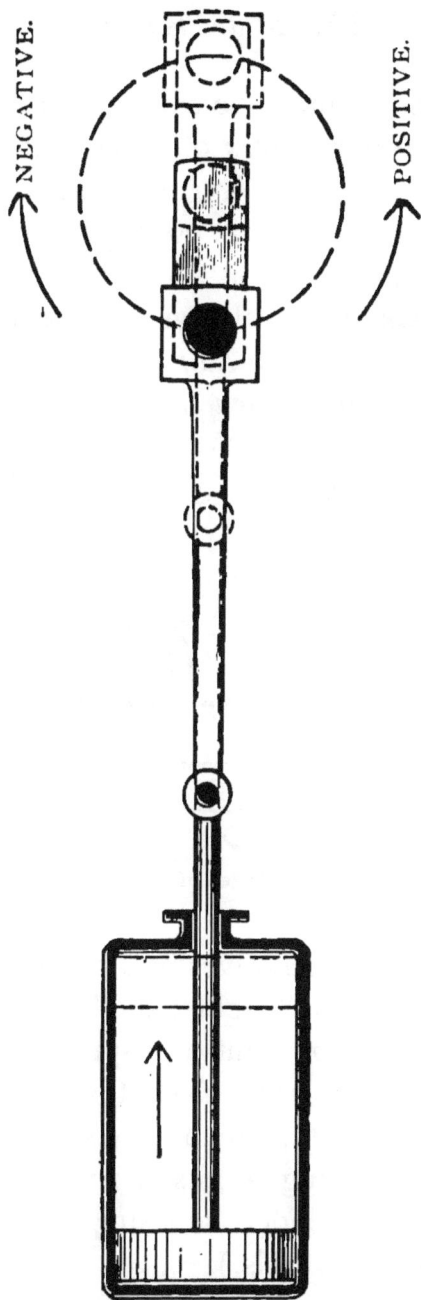

NEGATIVE.

POSITIVE.

FIG. 11.—Dead Centre—Connecting Rod and Crank in Line.

that neighbour might be said to lead the other by being *three-quarters* of a revolution in advance (or *more* than *half* a revolution, where the cranks are not set at 90 degs.), and this is where a little confusion sometimes arises; the leading crank must always be taken as that one which is *less than half a revolution in advance of its fellow when the engine is running*

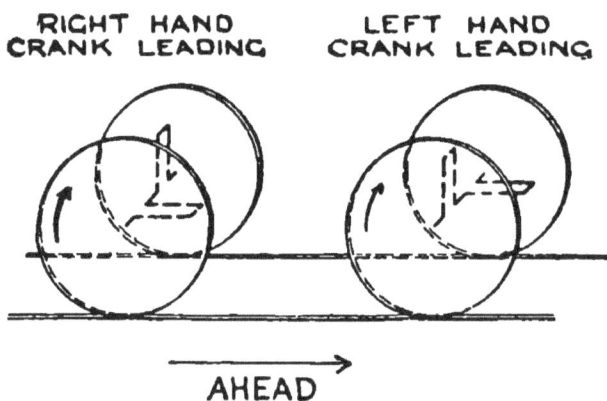

FIG. 12.—Order of Cranks.

ahead, and in locomotives and similar engines is " left-hand " or " right-hand," according as it lies to the left or right of a spectator looking from behind the crank-shaft towards the cylinders. (See fig. 12, which shows two arrangements of the driving cranks of an English locomotive. The cylinders are supposed to be to the right of the wheels.)

The accompanying diagram (fig. 13) will perhaps serve to make clear without further explanation the meaning of the expression, used
with reference to three-crank marine engines,
of " order of cranks, high, intermediate, low,"
or " order of cranks, high, low, intermediate,"
as the case may be.

FIG. 13.—Order of Cranks.

Let us revert here to one of the points previously noted, which was that the simple valve,
when we had it in connection with an eccentric
set 90 degs. in advance of the crank, would not
admit steam to the cylinder at the exact commencement of the stroke, *i.e.,* when the crank
is on either of the " dead centres."

Keeping this fact in mind, we will consider the matters of   .

Cushioning and Lead.—It has long been found desirable (especially in quick-running engines) that the motion of the returning piston should be opposed as it arrives at either end of its stroke, so that the moving weights of piston, piston-rod, and connecting-rod may be "cushioned," to prevent injurious stresses and to conduce to easy running. To use a homely illustration, one might say that just in the same way is it desirable, when striking out from the shoulder, to have one's adversary within range, that he may provide the necessary "cushioning," and so prevent stresses in one's arm and possible dislocations.

This "cushioning" is provided in part by checking the exit of the exhaust steam, as will be shown later, and in part by setting the eccentric a little more than 90 degs. (fig. 14) in advance of the crank, so that the valve commences to open the port to admit steam in front of the returning piston before it arrives at the end of its stroke, continuing to open the port as the piston comes to a rest, and is started on the return stroke.

The amount to which the port is found to be open when the piston is at the end of its stroke

is called the " lead " of the valve, and it would
be expressed as " one-eighth of an inch lead,"
or " three-sixteenths of an inch lead," etc.,

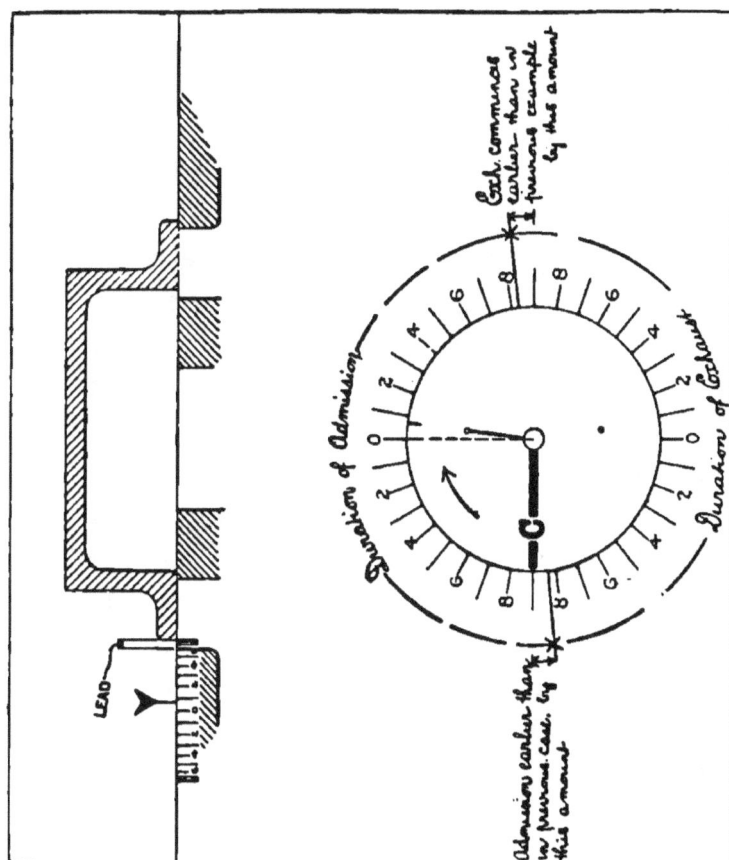

FIG. 14.

as the case might be. To set this valve for
lead, keep the piston at the end of its stroke,
and set the eccentric back until the valve is
opened to the desired lead; then the eccentric

would be fixed in its newly-found position, as shown in fig. 14. The radius of the " eccentric-arm," which is the lighter of the two radial lines on the disc in fig. 14, is equal in length to one-half of the travel of the valve, and therefore does not reach the edge of the disc, at which, however, there is a short guiding-mark in the line of the eccentric-arm " produced." By using this guiding-mark and moving the arrow-head of the valve over the scale on the ports, in correspondence with the movement of the guiding-mark on the edge of the disc within the circular scale, the valve, the main crank (C), and the eccentric can be operated in concord, very much as if they were mechanically connected.

The explanation just given of the meaning of lead, and of its effect, should be borne in mind, as we shall presently have to consider it in conjunction with other matters, for, as has been said, there is something more than " lead " employed in the production of cushioning.

CHAPTER V.

EXPANSION—INSIDE AND OUTSIDE LAP AND
LEAD; ADVANCE AFFECTED THEREBY—
COMPRESSION.

Expansion.—In order that the expansive properties of steam may be utilised, it must be cut
off before the piston arrives at the end of its
stroke, and on being cut off *must not be allowed to commence exhausting immediately,*
but must be confined in the cylinder, in order
that by expanding it may force the piston
further before it. The old valve, as we have
seen it, did not allow this, and to effect it the
faces of the valve are lengthened, so that after
the steam is cut off by the edge (*a*) of the
valve (fig. 15) the steam is confined, and expands in the cylinder until the edge(*b*) arrives
at *c*, when exhaust immediately commences.
Now put the piston at the end of its stroke,
and put the new valve in a central position
(fig. 16). The distance that the valve, when
in this central position, *overlaps the outer edge*

of the steam-port, is called the " outside or steam lap," or more commonly the " lap " of the valve, that is, the distance *d a* (fig. 16).

The first thing noticeable is that this valve must be given considerably more travel than the old one, as, to open fully each port to steam, we have to bring the edge *a* to *c*, first for one end of the valve and then for the other,

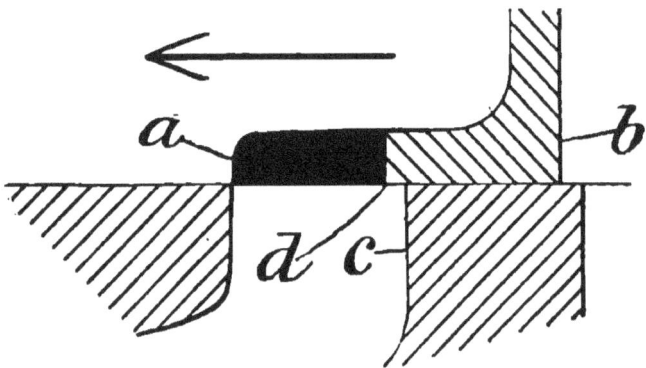

FIG. 15.

so that the travel of this valve must be *twice the distance from a to c*—i.e. (lap *plus* port-opening) × 2, whereas for the old valve it was the distance from *d* to *c* which had to be multiplied by two to give the travel.

Increased travel means increased throw of the eccentric, and we draw on the disc in fig. 16 two longer eccentric-arms than before (the bolder dotted lines) for the forward and backward eccentrics. We want "lead" with this valve, and we must therefore give each eccentric more advance than the 90 degs.

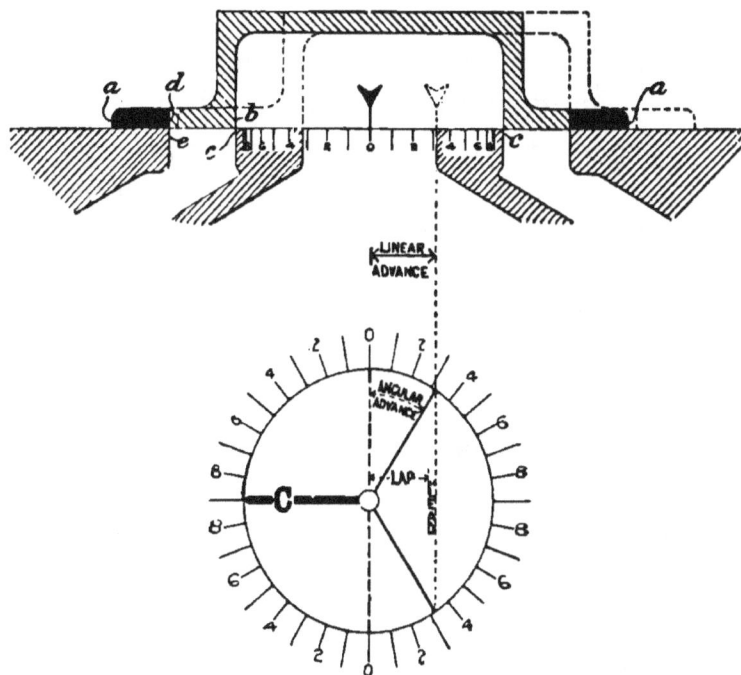

FIG. 16.

shown. Now with the old valve we give each eccentric an advance upon the 90 degs. which was determined solely by the lead, no lap being then in question; but we now have a lap to deal with, therefore each eccentric must be set on past 90 degs. until it first has drawn edge *a* of the lap to the outer edge of the port, and then still further until the edge *a* gives the necessary "lead." Hence we see that the linear advance of the eccentric = *lap plus lead,* so as to draw the valve out of its central position by that amount.

Travel, then, depends upon *outside lap and port-opening,* and lead does not affect it.

Advance depends upon *lap and lead.*

The advance of the eccentric is usually stated by specifying the angle of advance or the number of degrees *exceeding 90 degs.* that the eccentric is in advance of the crank—for instance, "an angle of advance of 10 degs." means 90 degs. + 10 degs. = 100 degs. in advance of the crank.

The line of travel of the valve is sometimes inclined to the line of travel of the piston; in such cases, one of which is illustrated in fig. 17, the angle of advance must, of course, be measured from a line at right angles to the valve-travel line, and not at right angles to the line of piston-travel, as we have hitherto assumed.

We must now carefully consider the further results obtained by the use of a valve having lap and giving lead, and whose eccentric of *greater throw* is advanced accordingly.

We should expect cut-off to take place

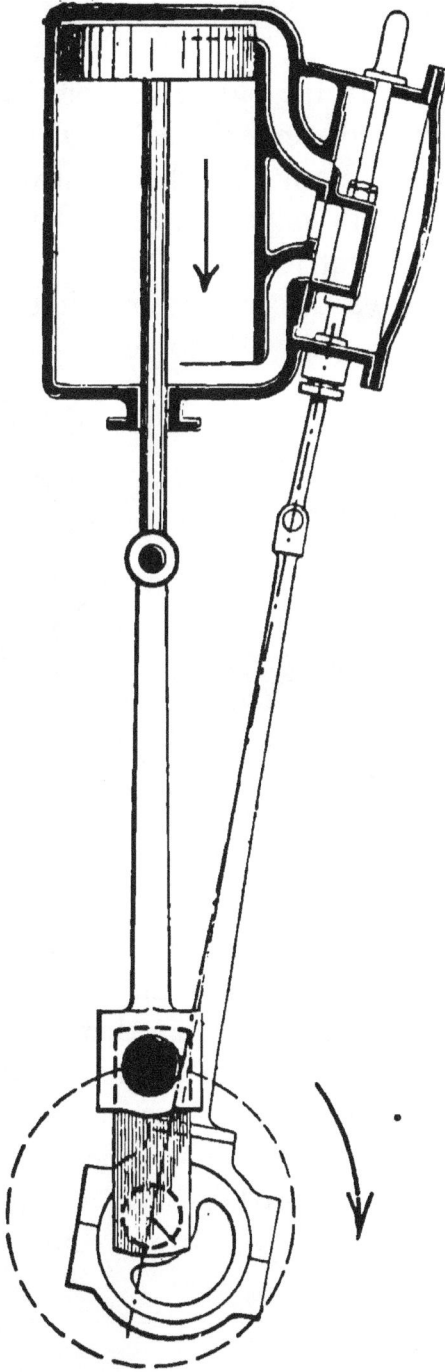

FIG. 17.—Engine in which the Line of Travel of Valve is Inclined in Relation to that of the Piston.

earlier than with the old valve, for the eccentric is advanced, and the edge *a* (fig. 15) would arrive at *e* a little earlier than would the edge *d* of the old valve, even without advance of the new eccentric.

This seen, we can understand that in causing the edge *a* (fig. 18) to reach the edge *e*

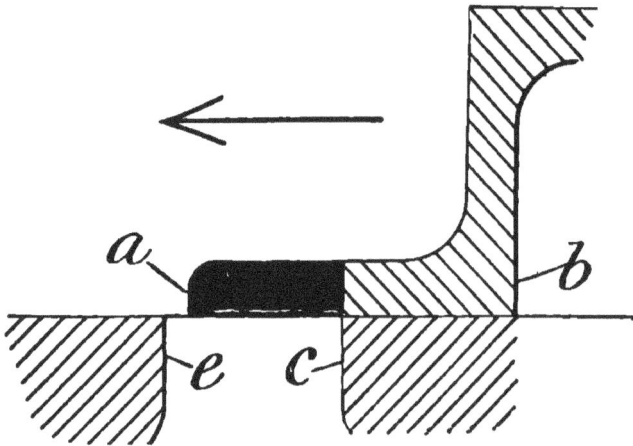

FIG. 18.

earlier in the stroke of the piston than it otherwise would do, we shall have hastened all the other. operations so that they must occur earlier in the stroke (not necessarily by exactly the same amount for each, however, as we shall see further on), hence the edge *b* (fig. 18) arriving earlier at *c*, exhaust will commence before the piston completes its

stroke, and also in returning to the *right*
(fig. 19) the valve is earlier, and instead of
the exhaust lasting during the whole of the
return stroke some of it will be shut in by *b*
and compressed before the returning piston
until the edge *a* arrives at *c*, whereupon the
lead commences coming in on and reinforcing

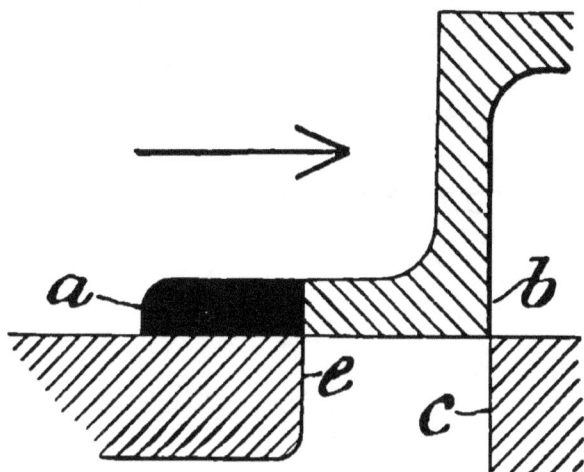

FIG. 19.

the compressed steam. Thus it is that compression and lead *together* affect the cushioning, and consequently the piston is eased in
stopping, and the return stroke is commenced
without injurious shock or jar. A similar
operation, of course, takes place at the opposite
end of the cylinder when the piston arrives
there.

The distribution-diagram for a valve with lap, lead, and a suitable eccentric is given in fig. 20, which should be compared carefully with figs. 10 and 14.

Inside Lap, exhaust lap, or " cover " as it is

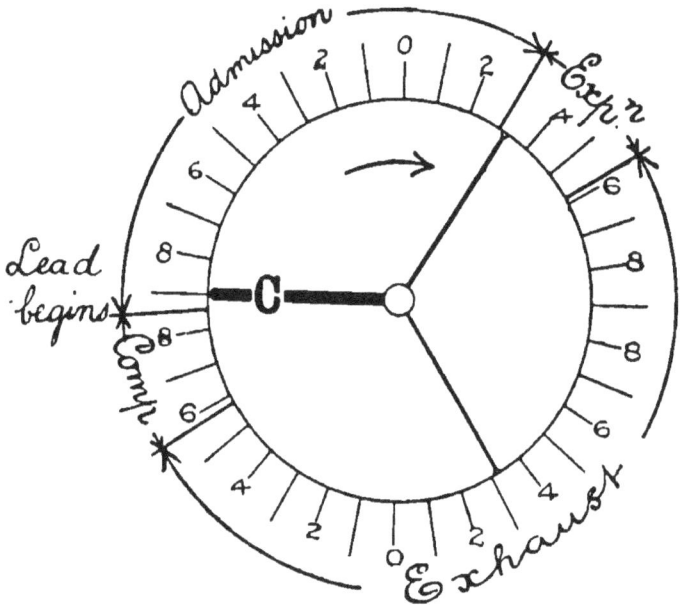

FIG. 20.

sometimes called, is possessed by many slide valves, and it is the amount that the valve when in its central position overlaps the *inner* edges of the steam-ports. Conversely,

Inside Lead is the amount of space left, in certain cases, between the inner edge of the valve and the inner edge of the steam-port when the valve is in its central position.

The amount of inside lap does not affect the setting of the eccentric, neither does it necessitate any alteration in the travel of the valve. With a given cut-off, steam is confined longer in the cylinder by a valve having inside lap because of the lengthening of the valve-face due to the inside lap, and compression will commence earlier and will be of longer duration from the same cause; exhaust being shortened in two ways—by the amount added to the duration of expansion at its end, and by a similar amount added to the compression at the commencement thereof.

Inside lead may be considered as something taken off the exhaust edge of the valve, and, in effect, exactly the opposite to exhaust lap, for, as may easily be seen, it shortens expansion and compression by reason of the shortening of the valve-face, and lengthens the exhaust at commencement and end by the amounts by which the operations of expansion and compression are shortened. The left-hand side of the valve in fig. 21 has inside lap, whilst the right-hand side has inside lead, and the circular distribution-diagram below compares the distribution in the case of valves having inside lap and inside lead with the distribution for an ordinary valve such as was considered with reference to fig. 20. By reference to the

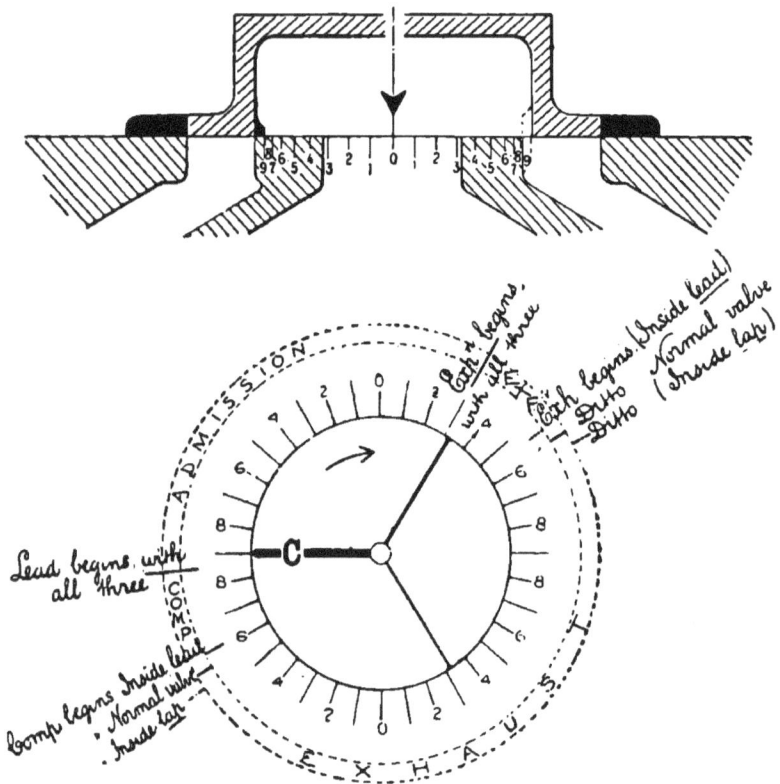

FIG. 21.

lower portion of fig. 21, it may be seen that the operations of admission and cut-off are unaffected, as they are performed by the outer edges of the valve; but

Expansion
{
is shorter and exhaust earlier with inside lead.

is longer and exhaust later with inside lap.
}

Compression
{
is later and shorter with inside lead.

is earlier and longer with inside lap.
}

From this it will be obvious why many ex-

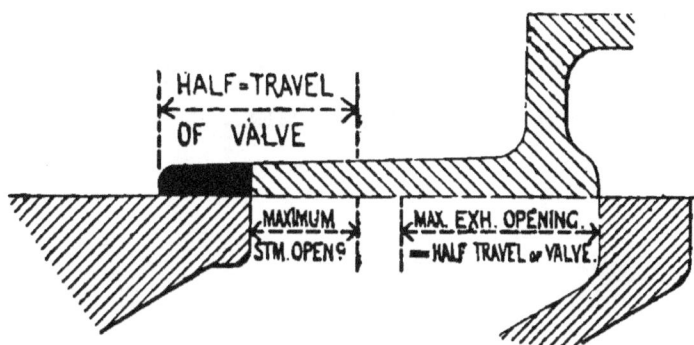

HALF-TRAVEL OF VALVE

MAXIMUM STM. OPENS

MAX. EXH. OPENING. = HALF TRAVEL or VALVE.

FIG. 22.

press locomotives are given inside lead, and why many good engines have inside lap.

Free Exhaust.—In order to give as free a passage as possible to the exhaust steam, so that it may leave the cylinder easily, and not have to be forced out by the returning piston and cause back pressure thereupon, it is customary to make the steam-ports wider than

the amount to which they are open to steam.
*This must on no account be confused with
inside lead,* which is quite a distinct matter.
The effect of having wide ports is that the
exhaust may be opened to any extent up to the
half-travel of the valve, although the steam-
opening may be much smaller, as may be
understood from the exaggerated diagram
(fig. 22), whence also should be seen that this
widening of the port does not affect the *duration* of expansion or of any other operation
of the valve, neither does it necessitate any
allowance in the setting of the eccentric. It
simply provides a free exhaust.

CHAPTER VI.

DOUBLE-PORTED, MULTIPLE ADMISSION, AND PISTON VALVES.

Double-ported Valves.—In large marine engines using considerable quantities of steam per stroke, and consequently requiring a large amount of port-opening, the ordinary slide valve would require an undue amount of travel. To obviate this, each end of the cylinder may have two steam-ports, both ports opening into one passage as shown, and a double-ported valve. The double-ported valve is virtually a combination of two slide valves in a single casting (fig. 23), one valve for the inner set of steam-ports and the exhaust-port, the other valve being halved, as it were, for each outer steam-port. The exhaust cavity of each of the outer " half-valves " having a passage leading to the exhaust cavity of the inner valve, both the valves exhaust into the central port, which is widened accordingly. The fig. 23 is merely diagrammatic, to enable the

principle of this valve to be easily borne in
mind; the next drawing (fig. 24) shows in
section a double-ported valve as actually constructed, with the exception that the valve-
spindle and its chamber are omitted in order
that the form of the valve may be more clearly
shown.   The section is taken in two planes,
one upon the longitudinal centre line of the
valve, and the other (the left-hand half of
the valve) a little to the near side of the centre

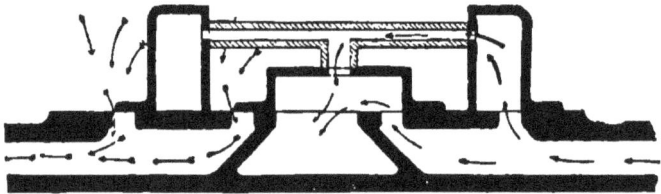

FIG. 23.—Diagram showing Flow of Steam and Exhaust
through a Double-ported Slide Valve.

line.   Fig. 25 shows a sectional view of a
double-ported valve in perspective.

Now, any movement given to the " inner
valve " is given, of course, to the outer one,
so that if the inner valve be moved so as to
give, say, 2 ins. of port-opening, the outer
valve gives at the same time another 2 ins. of
port-opening; therefore, with a given movement of a double-ported valve we get double
the amount of port-opening that we should

have obtained with the ordinary valve. Hence the travel of the double-ported valve = (lap + half the two port-openings) × 2; whereas that of the ordinary valve = (lap + the port-opening required) × 2.

The inner and outer valves of the double-ported construction, acting simultaneously, have each the same amount of lap; but they give the lead between them, and the eccentric is set with a linear advance equal to *one* of the

FIG. 24.—Vertical Section through a Double-ported
Slide Valve.

laps and *one* of the lead-openings. The distribution of steam with this valve is similar to that of the ordinary valve whose distribution-diagram was shown in fig. 20. The different periods are of the same duration in the example chosen, but the operations of cutting off steam and of opening exhaust are performed by the edges of the inner and outer valves instead of by a single edge as in the ordinary valve.

A valve very similar to the double-ported valve, shown in fig. 23, is the Giddings valve used on the Russell engines and some other automatic high-spced engines in America. By its peculiar construction, this valve obviates the necessity of double steam ports, yet it is so much like the double-ported valve that it may be noticed here rather than under the head of multiple admission valves, where it might be said to properly belong. This valve accom-

FIG. 25.—Perspective view (in Section) of a Double-ported Slide Valve.

plishes exactly what the double-ported valve does, and in pretty much the same way. The Giddings valve in fig. 26 shows steam being admitted to the right-hand end of the cylinder and exhausted from the left-hand end. The travel of this valve is, as in the case of the double-ported valve = (lap + half the port opening required) $\times$ 2. The valve should be set with a linear advance equal to the lap plus one-half the required lead.

MULTIPLE ADMISSION VALVES.

In America, the type of engine known as the automatic high-speed engine is very much used. These engines have a wide range of cut-off, from almost o to $\frac{1}{2}$ stroke, and in order to avoid a long valve and a long travel and to secure a quick admission and cut-off of the steam, recourse is had to multiple admission

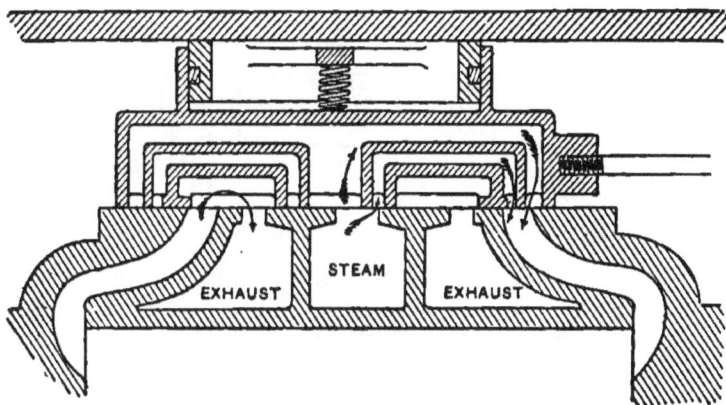

FIG. 26.—The Giddings Valve.

valves. These valves are so constructed that when one of them is moved from midposition, a distance equal to, say, 1-16 of an inch more than the steam lap, the width of the opening for the steam to pass through into the cylinder is 1-16 multiplied by the number of admissions of the valve. Thus, if the valve is a *double* admission valve, and it is moved 1-16 plus the steam lap from midposition, the port is opened a distance equal to $2 \times 1\text{-}16 = \frac{1}{8}$ of an inch. Double admission valves are quite common and there are also a number of quadruple admission valves used. They are made as flat valves and as piston valves, but all are made with a common object in view, that is, to secure an early cut-off and quick opening and closing of the ports with a valve whose length and travel shall be short.

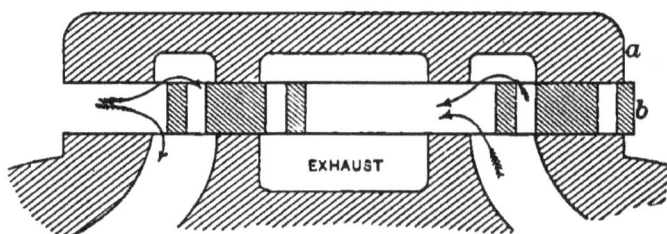

FIG. 27 --The Straight Line Valve.

One of the best-known forms of double admission valves is the " Straight Line " or " Sweet " valve, which was invented by Professor John E. Sweet, formerly of Cornell University, and first used by him on the Straight Line engines. This valve is now used by a large number of manufacturers of automatic high-speed engines. The valve is shown in section in fig. 27, where *a* is the " cover plate," and *b* is the valve, proper. The valve, *b,* moves back and forth between the cover plate, *a,* and the valve seat. The cover plate rests on pieces called " distance pieces," placed at the sides of the valve. These distance pieces are made of such a thickness that while the cover plate cannot touch the valve, it is so close to it as to make a steam tight joint between the plate and the valve. Therefore, the valve acts as a piston of rectangular cross-section, moving back and forth between the cover plate and the valve seat. The result of its peculiar construction is that a balanced flat valve is secured ; a valve which works with very little friction, as there is no surface upon which the steam can act so as to press the valve against the valve seat. The cover plate is kept in place by the pressure of the steam against its back, aided by one or two small springs inserted between the plate and the steam-chest cover.

This valve should be set so that the distance the valve is from midposition when the piston is at the end of its stroke, or the linear advance, shall be equal to the steam lap plus one-half the desired lead. The travel of the valve is equal to twice the sum of the steam lap and one-half the maximum opening of the port for

steam, or 2 × (steam lap + ½ maximum opening of port for steam).

In the case of a quadruple admission valve,
such as the Woodbury valve shown in fig. 28,
the travel is equal to twice the sum of the steam
lap and one-fourth the maximum opening of
the port for steam, or 2 × (steam lap + ¼
maximum opening of port for steam).

The different phases of the movement of a

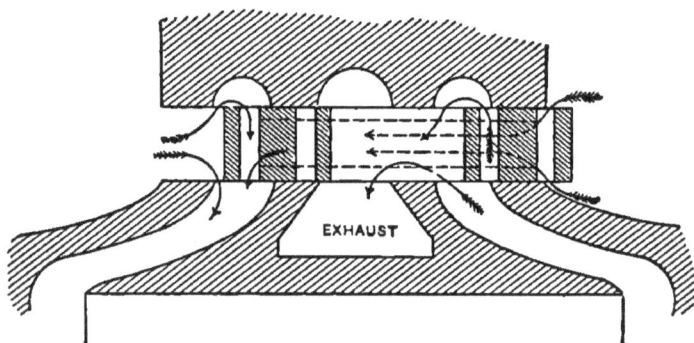

EXHAUST

FIG. 28.—The Woodbury Valve.

multiple admission valve are just the same as
for the ordinary slide valve, the only difference
being in the suddenness with which the ports
are opened and closed.

Piston Valves.—It may be readily seen that
between the ordinary flat slide valve and the
valve-seat considerable friction is set up, because the steam pressure on the back of the

valve is only partly balanced by the varying pressure in the cylinder steam-passages acting upon small portions of the area of the " rubbing " or under-side of the valve; but by making the port-faces cylindrical and making the faces of the valve in the form of pistons, the pressure is given no surface upon which to press in the direction of the valve-seat; or, in the case of hollow piston valves (an example of which will be presently illustrated), although the steam has access to the interior of the tube, pressure on parts thereof is balanced by an equal pressure in an opposite direction upon exactly opposite parts, and therefore is confined to stressing the valve and has no tendency to force any part of it on to the valve-seat.

In the plain piston valve shown (fig. 29) the faces are formed by the two valve pistons, which work in a cylinder wherein steam and exhaust openings are made.

Sometimes the steam-supply enters *between* the two pistons, as in figs. 30 and 31, exhaust taking place into the spaces at the ends of the valve, which spaces are connected by passages formed in the valve, which is tubular, as shown. In such valves the steam-lap must be put upon the *inner* edges of the valve pistons, for the steam and exhaust edges have their positions reversed; and the lead will also be given by the inner edges.

The eccentrics for a valve with internal

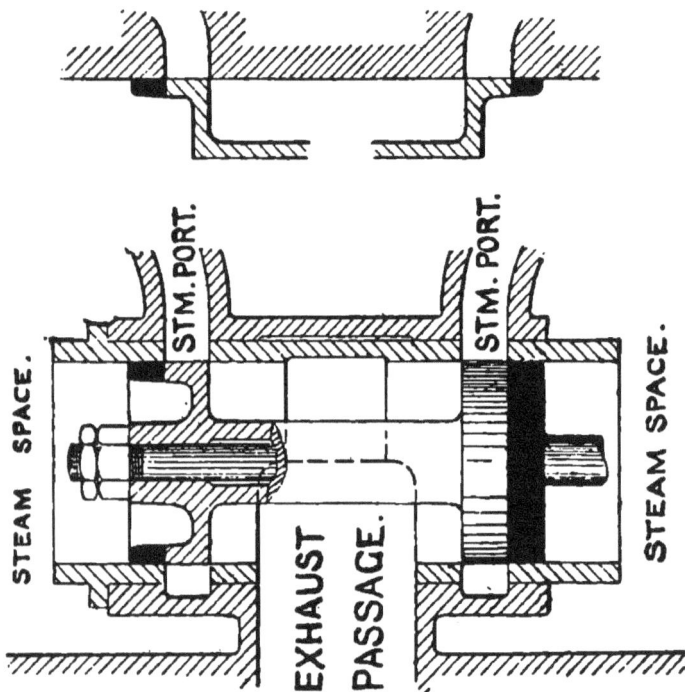

FIG. 29.

steam-supply must be placed 180 degs. in advance of the positions which they would occupy if the valve were of similar proportions, but with *external* steam-supply, for the reason that the internally supplied valve must be moving *up,* for instance, when the ordinary valve would be moving *down,* and *vice versa.*

If the valve be an externally supplied valve, but driven from one end of a two-armed rocking-shaft, the valve gear actuating the other end, as in American locomotive practice, the eccentrics must, for obvious reasons, be set 180 degs. in advance of their usual position.

Fig. 30 is given by the kind permission of Messrs. W. Denny & Company, the well-known Clyde ship-builders, and it possesses many points of interest, one of which is that the packing in the valve-spindle stuffing-box below the valve, and not shown in the drawing, is subject only to the influence of the temperature and pressure of exhaust steam, instead of to that of the higher temperature and pressure of the supply steam; this is a matter of importance, having in view the high pressures of steam which are now used.

Further, it may be seen that the upper piston of the valve is of greater area than the lower one, and that the high-pressure steam entering between the two pistons will consequently exert a total pressure upward upon the larger piston greater in amount than the downward pressure exerted upon the lower

and smaller one; thus the valve has a tendency
to rise, so long as steam is on. This upward
tendency balances wholly or in part the down-

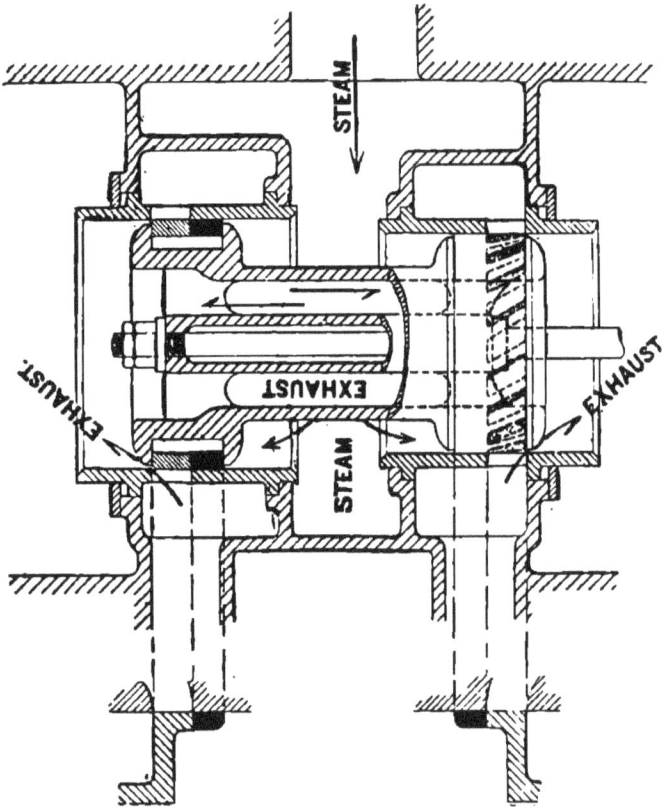

FIG. 30.

ward tendency due to the weight of the valve,
which has not to be lifted by the valve gear,
which may be of a lighter character than
usual, and give less trouble in working, owing

to diminution of wear and tear. The valve-spindle of the ordinary flat valve is sometimes prolonged (where the valve works vertically) to end in a small piston exposed to steam pressure, and acting so as to balance the weight of the valve.

In the " Joy " assistant-cylinder, a balance-piston is provided with special passages which,

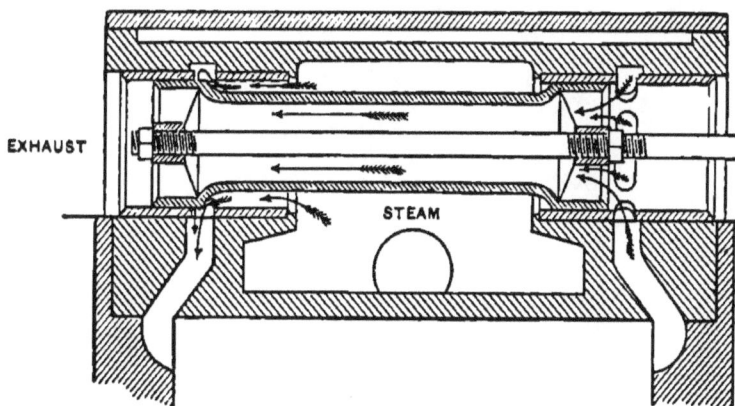

FIG. 31.—The Ide Valve.

in the movement of the valve-spindle, are brought opposite ports in the assistant-cylinder wall, so that steam is let in to propel the balance-piston up or down, and thereby assist the valve gear in each stroke. The balance-piston acts as its own exhaust valve in sliding over exhaust-ports in the assistant-cylinder wall.

Fig. 31 shows the piston valve used in America on the well-known Ide and Ideal engines. It is very similar to the valve shown in fig. 30, and has the same advantages which have been pointed out as belonging to that valve. As, however, it is principally used on horizontal engines, the two pistons are made of the same diameter, in order to secure a perfectly balanced valve.

CHAPTER VII.

THE EFFECT OF ALTERATIONS TO VALVE AND ECCENTRIC.

Advance of the Eccentric.—It should here be established that to any given example of valve, ports, and eccentric (leaving the piston out of consideration for a moment) belongs a certain cycle of operations, the occurrence of these operations and their *duration, expressed in fractions of a revolution of the rotating eccentric,* being constant for a given case,* but the fraction of stroke traversed by the *piston* between or during any of these operations of the valve will depend upon *where* in the stroke of the piston the cycle of operations is started, taking any convenient operation as

---

* To make the idea clear one might plot these operations round a circle representing one revolution of the eccentric, when it would be seen that the operations of the *valve* could be considered quite independently of those of the piston—then it would remain to show what the piston might do, or have done upon it, by comparing its motion under varying conditions with the *unvarying* operations of the valve, in the manner explained later.

being the commencement of the cycle, for the
speed of the piston rises from *nil* at the commencement of its stroke to a maximum near
the middle, falling again to *nil* at the end.

Now we may go on to consider the effect
of increasing the advance of an eccentric beyond that which it would ordinarily have
(fig. 32). If we suppose our eccentric and
valve to give us a cut-off at C and an exhaust
at E, expansion lasting 0.17 of the stroke (as
shown in full lines, fig. 32), and that by advancing the eccentric we get a cut-off at C^1
0.14 earlier, we must not expect to find that
exhaust also will be 0.14 earlier (dotted lines,
fig. 32), for although the exhaust edge of the
valve always gives exhaust at a certain constant period of time later than steam is cut
off by the steam edge, yet in that interval the
speed of the piston slackens less in this second case than in the first example, and it
therefore travels through a greater distance
than in the same time-interval taken later in
the stroke in the first case. Hence, with increased advance we get *earlier* cut-off, with
expansion of *lengthened duration* because *exhaust is not earlier in the same degree,* neither
are all the other operations performed by the
valve, although they each commence earlier

and last longer in the stroke, lead included, but admission, of course, excepted. *The idea may, perhaps, be more easily grasped by keep-*

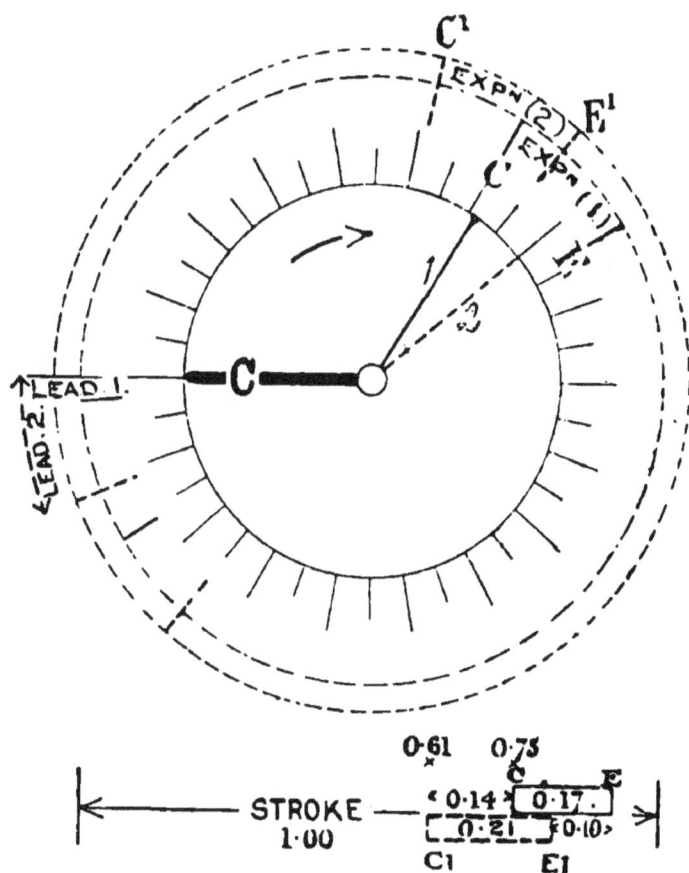

FIG. 32.

ing the eccentric and valve in view at an un-altering standard of reference, and when it is desired to consider the effect of advancing the

*eccentric, to imagine instead that it is the set-
ting of the main crank which is altered—*
which in effect is the same thing, and is, perhaps, easier to reason out.

*Lengthening or Shortening the Eccentric
Rod—Shifting the Valve on its Spindle.—*
Starting with the valve in its central position,
and lengthening or shortening the eccentric
rod or shifting the valve on its spindle, or
combining both operations in the same sense,
the distance between the centre of the crank-
shaft and the centre of the valve is altered,
and the travel of the valve is shifted bodily to
the right or left, *i.e.,* the valve when in the
centre of its travel (the extent of travel remaining of course unaltered) will be either
further from or nearer to the crank-shaft than
before, while the position of the ports remains
unaltered; therefore the arrow upon the valve
at its centre (see fig. 16) lies to the right or
left of the centre line of the port-face, and,
that arrow being no longer appropriate as a
pointer at mid-travel to the zero of the scale
on the ports, we must adopt as our index-
finger the graduation mark made adjacent to
the arrow and lying over the said zero, in
order that when the disc may indicate that
the valve should move, say, from o to 6 the

new mark or index-finger, and not the arrow, may be given that movement.

If the valve had previously been properly set for lead, and the alteration shifted it $\frac{1}{8}$ in. to the left, for example, we should find that the lead at the left-hand end would be diminished by that amount, and that subsequently, at cut-off, release, and compression, for both ends of the cylinder, the valve would always be found $\frac{1}{8}$ in. to the left of its proper position. The result would be, as shown in fig. 33, that: At left-hand of cylinder (see lines of dashes) :—

Lead decreases
Cut-off takes place
Exhaust commences

} earlier, these operations taking place as the valve moves *towards* the left.

Compression commences later, this operation taking place as the valve moves *from* the left.

At right-hand end of cylinder (see lines of dots) :—

Lead increases
Exhaust commences
Cut-off takes place

} later, because when effecting them the valve is moving *from* the left.

Compression commences earlier, for it occurs whilst the valve is moving *to* the left.

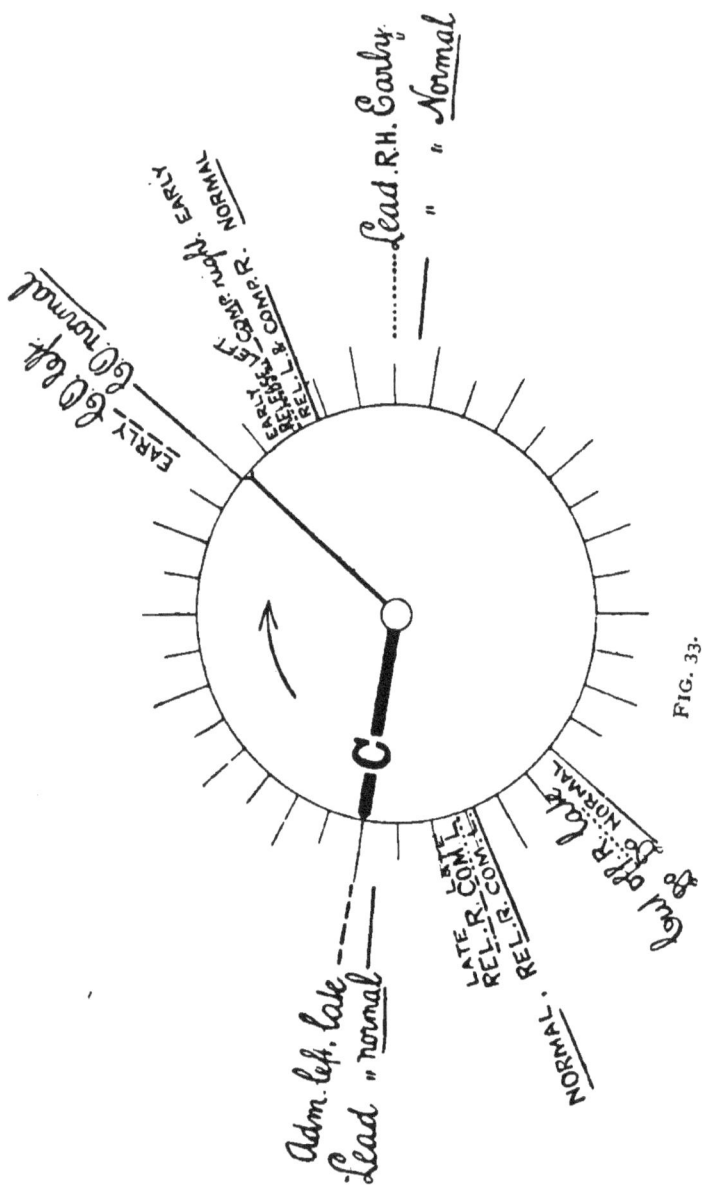

Lead R.H. Early.
" " Normal

NORMAL
EARLY COMP. R.
RELEASE & COMP. R.
REL. L & COMP. R.
EARLY

Adm. left. late
Lead " normal

EARLY Lead EX left
EX left normal

C

LATE COMP. L.
REL. R. COM. L.
REL. R. COM.

NORMAL.

Lead EX R left
NORMAL

FIG. 33.

The alterations at opposite ends being in opposite sense. (The normal distribution is shown by *full lines* for comparison.) And although the commencement of the operations is advanced and retarded by the same amount of error, operating in opposite sense on opposite sides of the piston, yet there is a difference in the work done upon the opposite sides because of the commencement of the two cycles of operation at different times, whence we infer differences in the speeds of the piston, in the quantities of steam admitted, and in the expansion-curves.

In some cases such a difference is desirable, and with the working diagram suggested in Chapter I. another series of results may be obtained by using the ordinary valve for distribution on one side of the piston, and for the other side using a differently-proportioned valve with a different index-pointer, as though opposite ends of one valve were differently proportioned, as in certain large vertical engines.

Adding Extra Lap to a Valve.—In some cases cut-off may be wanted earlier in the stroke, permanently, with expansion of longer duration. Under such circumstances more lap may be added to the valve. If this be done, and the valve with its extra lap be re-set by giving the eccentric greater advance in order to get the same lead as before (travel of valve remaining unaltered), the steam-ports will not be opened to so great an extent as formerly, and will close earlier, partly because of the added lap,* and partly because of the increased advance. The increased length of the valve-face will, after cut-off, confine the steam longer in the cylinder, and it should be obvious (the effect of advancing the eccentric having already been shown with reference to fig. 32) that in addition to the earlier commencement and lengthened duration of expansion, the exhaust and compression will begin somewhat earlier because of the increased advance, and the compression will be of longer duration because of the increased length of the valve-face.

General.—It is here seen that in altering either the outside or the inside laps with intent to affect only the steam admission with the

---

* To assure one's self of this, take an extreme case and imagine an excessive amount of lap to be added. In such a case, with travel unaltered, the ports might not be opened to steam at all, the valve simply sliding backwards and forwards without uncovering them ; now, taking this lap off piecemeal the port would first be opened to an exceedingly small degree and instantly shut again, the amuont and duration of opening increasing with each successive removal of lap.

former, or only the exhaust with the latter, we must *perforce* affect other things. For instance, we alter the outside lap, with the intention, say, of simply making cut-off earlier or later as the case may be—but having altered the lengths of the valve-faces, they will alter the duration of expansion and of compression, and the alteration necessitated in the advance of the eccentric in order that the lead may remain unaltered has its own effect in addition. This interdependence, which is often inconvenient, is characteristic of the slide valve worked by an eccentric, and has led to the use of valve gears and valves which permit, more or less, of the independent regulation of some or all of the various operations in the distribution of steam. Some of these gears are used with a slide valve (as in the case of the Joy and other well-known valve gears), while others operate special valves.

CHAPTER VIII.

The Link Motion.—In engines whose duty *frequently* varies, as in the familiar case of a locomotive, permanent alteration of lap would not meet the requirements of the case, for the amount of expansion should be susceptible of ready variation at any time, in accordance with alterations in the weight of train, gradients, and weather. The link motion, in addition to being a reversing gear, is also a "variable expansion" gear. A precise analysis of any given case of link motion is a somewhat difficult operation, but, speaking broadly, the link motion may be said to provide what is virtually an eccentric of different throw and advance for each degree of expansion. Although, actually, the same two eccentrics remain constantly in use, the link motion may be so adjusted as to have an effect on the valve *similar to that which would ensue*

upon the substitution, one for another, of different eccentrics as stated. Assuming that the
lowest range of expansion is being used, the
valve will have the travel, cut-off, and lead due
for the most part to the throw of the actual
eccentric and to its position with relation to
the crank, almost as in the case of an eccentric
whose rod is directly connected to the valve-
spindle; now, as the link is raised (an operation sometimes termed " linking-up ") its centre gets nearer to the block on the end of the
valve-spindle than it was before (*i.e.*, nearer
to " mid-gear "), and we get in effect:—

*With link mo-
tion having
"open" rods.*
{
A series of eccentrics of decreasing throw, but of
increasing angular and
linear advance.
}

*With link mo-
tion having
"crossed"
rods.*
{
A series of eccentrics of decreasing throw and of
increasing angular, but
decreasing linear ad-
vance; throw, therefore,
decreasing more rapidly
than in the case of the
link motion with " open "
rods.
}

Take the case of a partially linked-up mo-

tion with open rods; we shall find that because
the throw of the eccentric is lessened, the

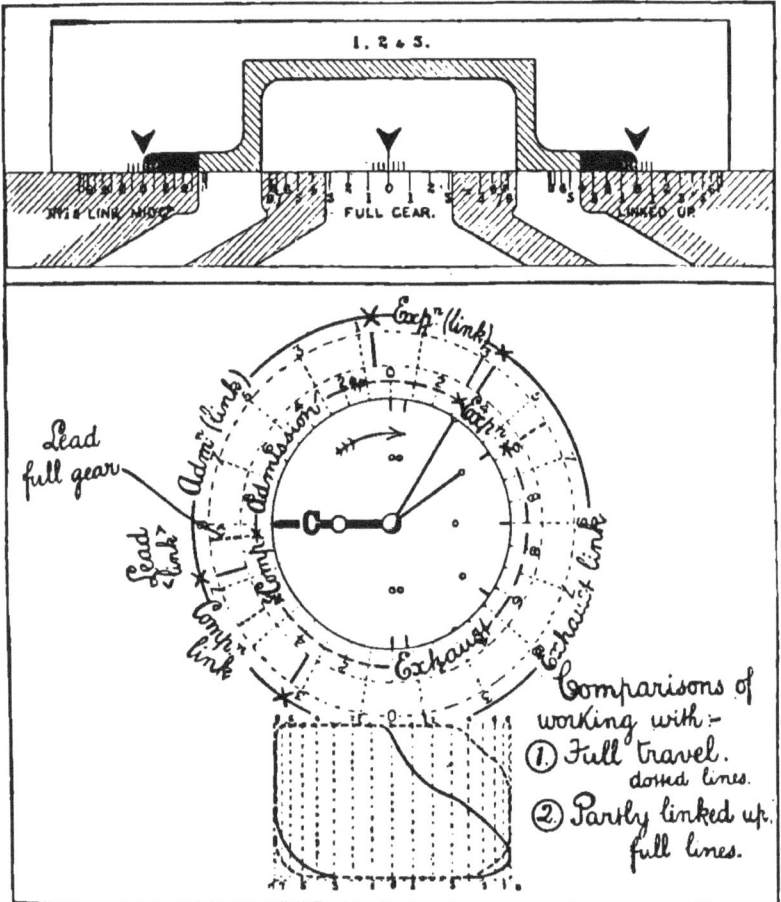

FIG. 34.

travel of the valve to the right and left of its
central position will be diminished, and the

ports will not be opened so wide to steam or exhaust; because of the increased advance of the eccentric, steam will be cut off earlier, compression and lead will also be earlier, expansion and compression lasting longer, and

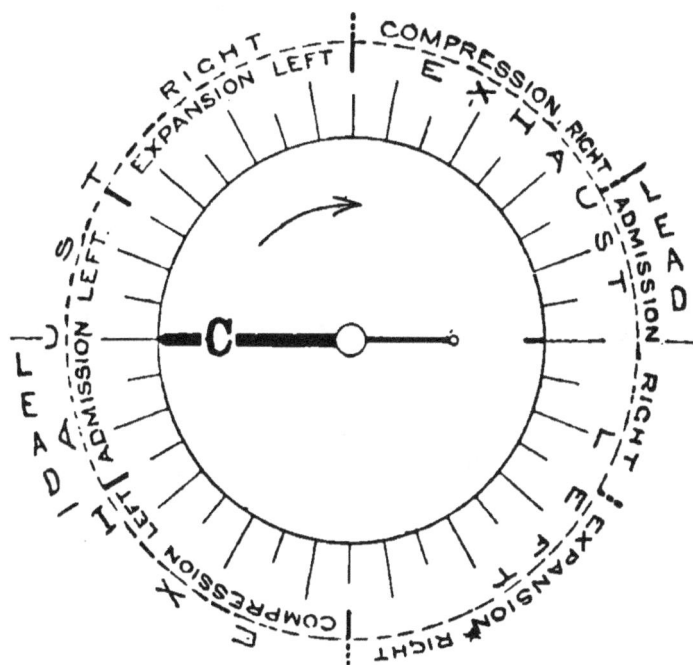

FIG. 35.

exhaust and admission being shortened. The difference between working in full gear and linked up is shown by the diagram in fig. 34.

Below are diagrams showing the effect of putting the link motion in "mid-gear" (fig.

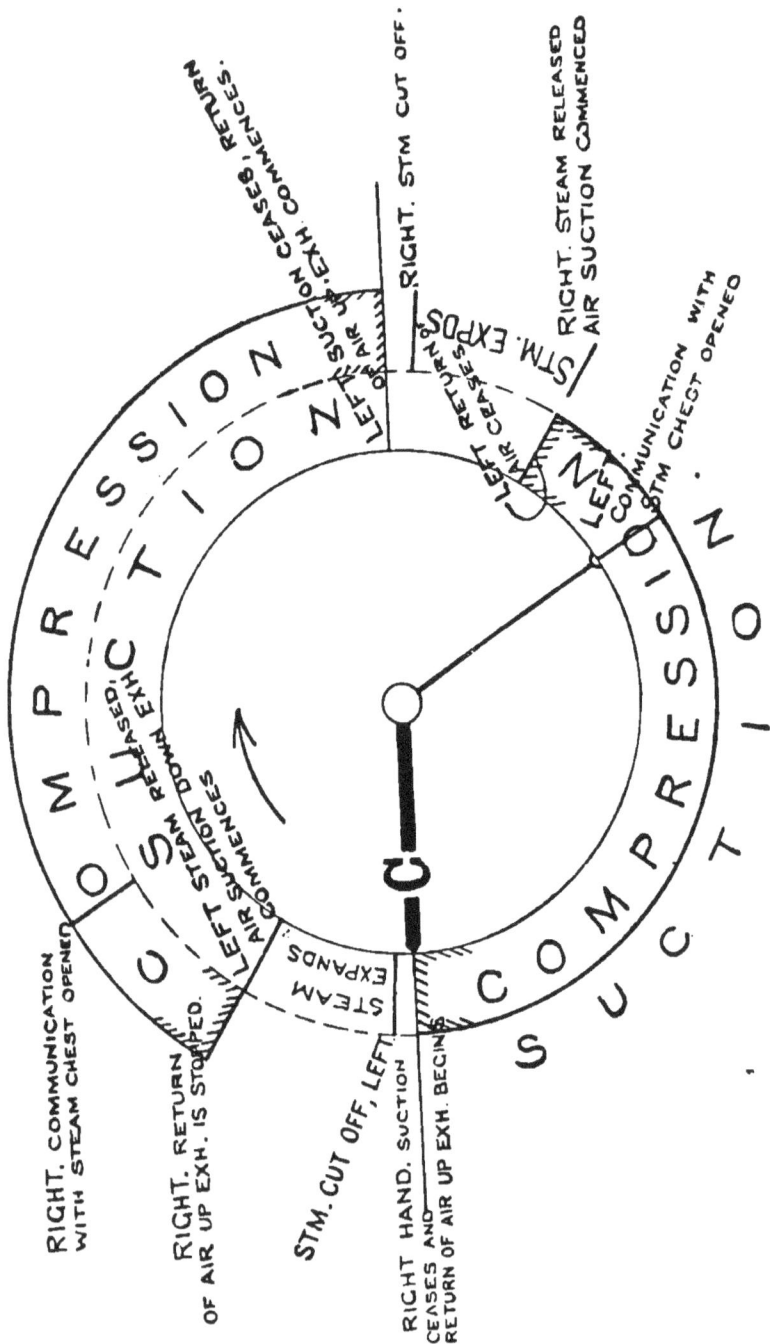

RIGHT. COMMUNICATION WITH STEAM CHEST OPENED

RIGHT. RETURN OF AIR UP EXH. IS STOPPED.

STM. CUT OFF, LEFT SUCTION

RIGHT HAND. SUCTION CEASES AND RETURN OF AIR UP EXH. BEGINS

RIGHT. STM CUT OFF.

RIGHT. STEAM RELEASED AIR SUCTION COMMENCED

LEFT. SUCTION CEASES, EXH COMMENCES.

LEFT. COMMUNICATION WITH STM. CHEST OPENED

COMPRESSION

EXH.

SUCTION

STEAM EXPANDS

COMPRESSION

SUCTION

STM. EXPDS.

LEFT RETURN OF AIR CEASES

RELEASED EXH.

LEFT STEAM DOWN AIR SUCTION COMMENCES

LEFT. AIR

FIG. 36.

C

C

35) and of putting the link motion into " back
gear " (fig. 36) while the engine is running

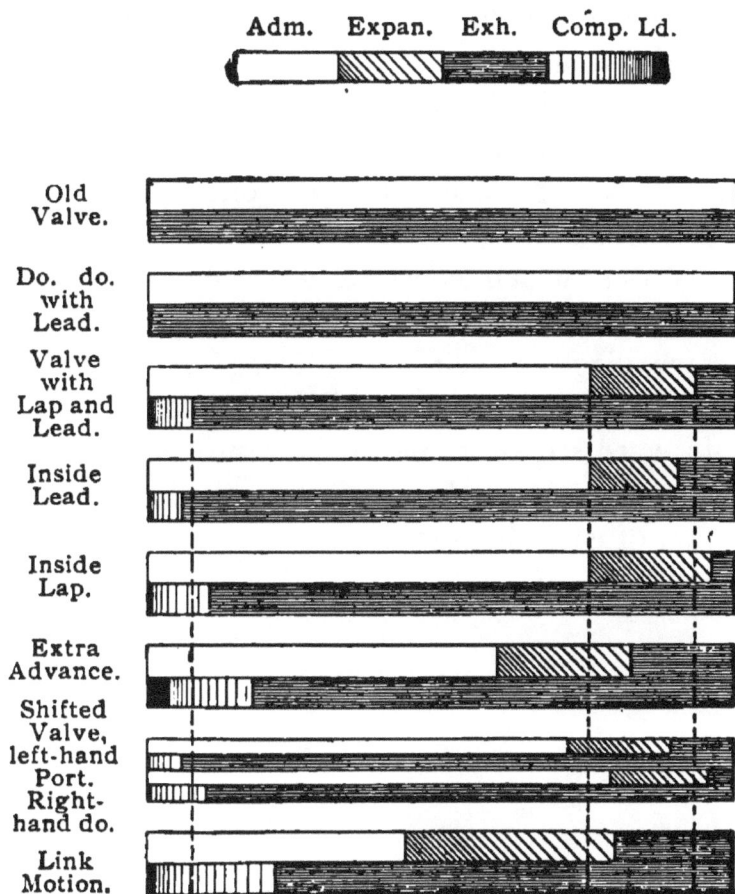

FIG. 37.

ahead, as for a locomotive, but it will be
understood that the effect of the compression
shown in the latter will be modified by the

lifting of the valve from its face, with the result that some compressed air and steam will go up the exhaust.

The set of diagrams forming fig. 37 is intended to give a comparative view of the effect upon the distribution of steam, of differently arranged valves and eccentrics. The operations taking place upon only one side of the piston are shown, except in that section of the diagram which concerns a shifted valve, in order to avoid confusion. The upper line of each diagram, being read from left to right, shows the operations which take place on the forward stroke, those of the return stroke being read from right to left upon the lower portion of each diagram.

CHAPTER IX.

Very early Cut-off.—Theoretically there is no
limit to the range of cut-off which may be obtained from an ordinary slide valve, but a
moment's consideration will show that if we
require a valve to cut off steam very early in
the stroke of the piston so as to cause it to ex-
pand throughout an increased fraction of that
stroke, the *lap* must be considerably increased
beyond that which is usual, if the valve-face
is to be enabled to keep the steam confined in
the cylinder for a sufficient period after cut-off
is effected.

As a consequence of increased lap, the *travel*
of the valve, made up of *lap plus the required
port-opening*, will need to be increased correspondingly ; the *linear advance* (the sum of lead
plus an *increased* lap) will be increased also,
and the valve will therefore come back earlier

into its central position (where compression begins) than would the normal valve, so making
compression begin earlier and last longer than
before. In an unbalanced valve, this increased
travel, together with the abnormal compression, cannot be tolerated, so that for obtaining
an early cut-off without too much compression, with a sliding valve, gear of the " separate cut-off slide," or Meyer type, is employed. This is characterised by a cut-off

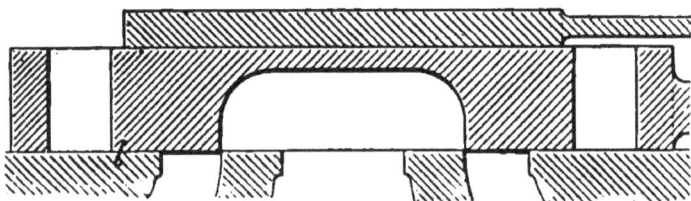

FIG. 38.—Slide Valve, with Cut-off Plate on back.

plate or plates on the back of a main slide
valve, which may have the usual exhaust
cavity, and in addition possesses supply steam-
conduits leading through the valve from the
back to the " lap-edges " (fig. 38). Exhaust
and compression are timed by the edges of the
central exhaust cavity; lead is given by the
" lap-edges," but cut-off is controlled by movement of the cut-off plates on the back of the
moving main valve, so that though a " lap-

edge " may not at any time have closed one of the steam-ports proper, a cut-off plate worked by a separate eccentric can close the appropriate steam-conduit above referred to, and so stop the current of steam flowing past the lap-edge to the still open port. Moreover, by a suitable setting of the eccentrics of the main and cut-off valves, it is possible to arrange that, at the instant of cut-off, the two valves shall be moving in opposite directions, and so shall effect a sharp cut-off, whereby wire-drawing is reduced.

With the model referred to in the earlier chapters, the action of a Meyer or similar gear can be readily investigated if an extra eccentric-arm be added to the crank-pin disc and an extra travel-scale marked upon the baseplate.

In fig. 39 is shown the " Buckeye " valve used in America on the Buckeye engine; *a* is the main valve, which determines the admission, release and compression of the steam; and *b* is the small auxiliary valve which rides on the top of the main valve *a,* and which effects the cut-off. This valve is constructed so as to have all the advantages of a perfectly balanced valve as well as those of the Meyer valve. There are two faces between which the

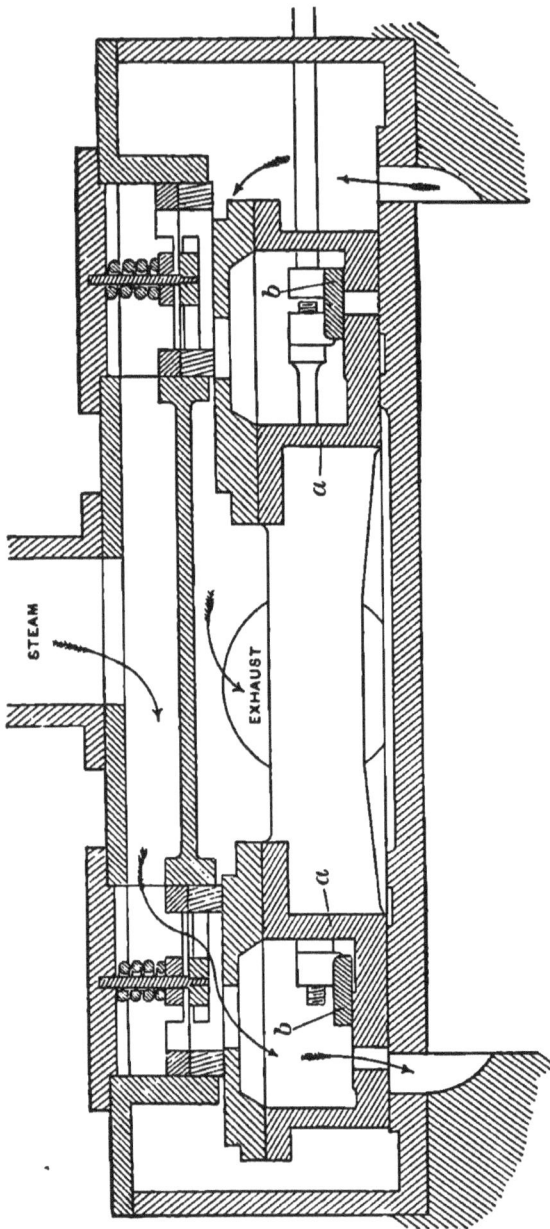

FIG. 39.—The Buckeye Valve.

main valve slides steam tight, and the small valve works on a face inside of the main valve, as shown in the figure. Steam is exhausted past the outer edges of the main valve. In order that steam may enter the cylinder, the port in the main valve must be in communication with the steam port and at the same time it must not be covered by the auxiliary valve *b*. In the figure the valves are shown in such a position as to admit steam to the left-hand end of the cylinder and to allow it to exhaust from the right-hand end, as indicated by the arrows.

Each valve is moved by its own eccentric, quite independently of the other. By changing the advance or eccentricity, or both, of the auxiliary valve, the cut-off may be changed as desired, without in any way affecting the motion of the main valve. And as the main valve determines the admission, release, and compression of the steam, it is possible, with this valve, to change the cut-off as desired without affecting these at all.

The Trick or Allen Valve.—In connection with the subject of early cut-offs the Trick or Allen valve may be mentioned. It is made with a passage-way through the back of the valve, as shown in fig. 40, adapted to deliver an extra supply of steam into the port by ren-

dering available an additional proportion of
the area thereof, when the narrow port-opening given by the lap-edge alone, during
linked-up working, is found to wire-draw the
steam. Whilst the lap-edge a, for instance,
uncovers the edge b of the right-hand port,
the left-hand end of the Trick passage becomes
uncovered by moving beyond the edge c of the
port-face, so that a current of steam flows
from left to right through that passage to join

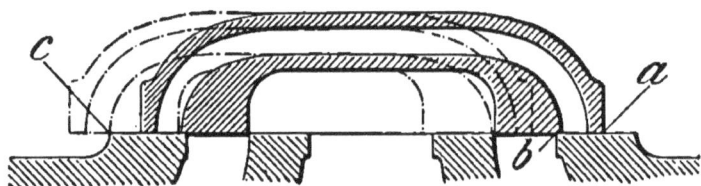

FIG. 40.

the stream passing the edge a. The total area
of the two supply-streams flowing into the
right-hand port being practically twice that
which would be available were the Trick passage absent.

The occurrence and duration of the operations of steam-distribution are exactly the same
with this valve as with an ordinary slide-valve
of corresponding proportions, the distinctive

feature of the valve having no effect upon them so far as their timing is concerned.

Reversal.—It may be of interest to consider the question—"Why do we need *reversing* gears?" The gears in common use are nearly all variable expansion gears, as well as being *reversing* gears, but it is desired to consider now only the question of reversing. Are reversing gears really needed? Suppose that one could have a transparent cylinder and valve-chest, so that movement of the piston and valve could be clearly seen. Suppose also that the connecting-rod, crank, valve gear, and fly-wheel could *not* be seen. We look into the cylinder and see valve and piston reciprocating regularly, and in harmony with one another. We turn away, and before we look again, the engine, let us assume, has been reversed without our knowledge; yet when we turn again to the engine there are the piston and valve reciprocating just as they were at first (*although they must have "changed step," as we shall presently find*), and, look as intently as we may, with nothing else visible we shall be unable to tell that the engine has been reversed. The valve is doing now exactly what it did before, performing its operations in the same order. Why have we had to employ reversing gear? Solely because—when the engine ceases to run *ahead* for instance, stopping at half-stroke the piston and valve will be in a certain position, and will have been moving in certain directions relatively to one another. And to reverse the *piston's* motion, we want the valve

Firstly—To take up a new position on the port-face, so as, for example, to give steam where it had previously given exhaust, and thus we get reversal of the *engine;* and

Secondly—To move away from that new position, after the engine has started astern, in the same direction as that in which it was moving when the engine was running ahead. And seeing that the valve gear derives motion from the crank-shaft directly or indirectly, we cannot give the valve motion in this *same* direction from an engine now moving reversely, unless we also reverse the *gear.* And thus we may conclude that a reversing gear is needed simply for the purpose of regulating the direction of the first motion of the valve after stopping (*i.e.,* for " changing the step "), for after it next reaches the end of its travel its motion will be in all respects just as it was when the engine ran ahead,—so that, if we had a single-cylinder engine and never stopped it except on the dead centres, we could drive its valve *by a separate valve-engine always running one way,* and our main engine would, if pushed off by hand, run equally well either ahead or astern, whilst served by a valve having lap, lead, and travel as is customary. It is hoped that this little speculation may not appear unprofitable, for it has a bearing on the action of the reversing slides in radial valve gears such as those of Hackworth, Walschaert, and Joy.

FIG. 41 SHOWS ENLARGED KEY-VIEW OF CARD-
   BOARD CRANK-DISC, WHICH SHOULD TURN
   WITHIN THE CIRCLE DRAWN IN THE CENTRE
   OF FIG. I.

The arm marked C is the crank-arm, and serves as an index finger in the recording and reading of results.

The mark 1 is the index-mark for the forward eccentric to operate a valve without lap.*

The mark 2 is the index for a forward eccentric set for lead, using a valve without lap.*

The mark 3 is for a forward eccentric for a valve with lap and lead run in " full gear."†

---

* Used with No. 1 travel-scale, fig. 1.
†   "   " No. 2   "   "   "

The mark 4 is for a forward eccentric for a valve with lap and lead run with a link motion partially "linked up."‡

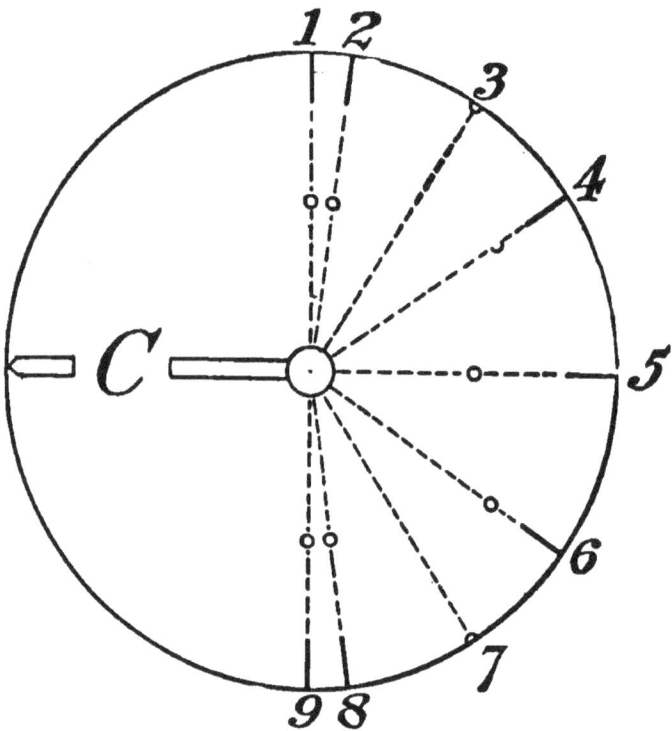

FIG. 41.

The mark 5 is for a valve with lap and lead run with the link motion in mid-gear.*

The mark 6 is for a backward eccentric for

‡ Used with No. 2 travel-scale, fig. 1.
*   "   .   "   No. 1   "   "   "

a valve with lap and lead, run with a link motion partially linked-up.‡

The mark 7 is a backward eccentric for a valve with lap and lead run in " full gear."†

The mark 8 is for a backward eccentric set for lead, using a valve without any lap.*

The mark 9 is for a backward eccentric for a valve without lap.*

The radial lines between the centre of the disc and the small circles represent the actual throw assumed for the various eccentrics and their angular position in relation to the crank; where the actual throw is not sufficient to enable the arms to reach the edge of the disc, the throw-lines are produced to the edge.

‡ Used with No. 2 travel-scale, fig. 1.
† "   " No. 2   "   "   "
* "   " No. 1   "   "   "

THE
FIREMAN'S GUIDE
A Handbook on the Care of Boilers
BY KARL P. DAHLSTROM, M.E.

CONTENTS OF CHAPTERS

8vo, cloth, 50 cents.

THE CORLISS ENGINE.

By John T. Henthorn,

—AND—

MANAGEMENT OF THE CORLISS ENGINE.

By Charles D. Thurber.

Uniform in One Volume. Cloth Cover; Price, $1.00.

Table of Contents.

Third Edition, with an Appendix.

HOW TO RUN
Engines and Boilers

Practical Instruction for Young Engineers and Steam Users.

BY EGBERT POMEROY WATSON

REVISED AND ENLARGED

Synopsis of Contents

Cleaning the boiler, removing scale, scale preventers, oil in boilers, braces and stays, mud drums and feed pipes, boiler fittings, grate bars and tubes, bridge walls, the slide valve, throttling engine, the piston, testing the slide valve with relation to the ports, defects of the slide valve, lap and lead, the pressure on a slide valve, stem connections to the valve, valves off their seats, valve stem guides, governors, running with the sun, eccentrics and connections, the crank pin, brass boxes, bearings on pins, adjustment of bearings, the valve and gearing, setting eccentrics, the actual operation, return crank motion, pounding, the connections, lining up engines, making joints, condensing engines, Torricelli's vacuum, proof of atmospheric pressure, pumps, no power in a vacuum, supporting a water column by the atmosphere, starting a new plant, the highest qualities demanded.

Water tube boilers, fire tube boilers, why water tube boilers steam rapidly, torpedo boat boilers, management of water tube boilers, economy and maintenance of water tube boilers.

150 pages, illustrated, 16mo, cloth, $1.00

THEORETICAL AND PRACTICAL

Ammonia Refrigeration

A Work of Reference for Engineers and others Employed in the Management of Ice and Refrigeration Machinery.

By ILTYD I. REDWOOD

CONTENTS

150 pages, 15 illustrations, cloth, $1.00.

LUBRICANTS,
OILS ❧ AND ❧ GREASES

**Treated Theoretically and Giving Practical Informa-
tion Regarding Their**

COMPOSITION, USES AND MANUFACTURE

BY ILTYD I. REDWOOD

CONTENTS

8vo, cloth, $1.50.

PRACTICAL HANDBOOK ON

Gas Engines

With Instructions for Care and Working of the Same,

BY G. LIECKFELD, C.E.

Translated with permission of the Author by

GEORGE RICHMOND, M.E.

WITH A CHAPTER ON OIL ENGINES

CONTENTS

120 pages, illustrated, 12mo, cloth, $1.00.

The Best and Cheapest in the Market

ALGEBRA SELF-TAUGHT

FOR THE USE OF

Mechanics, Young Engineers and Home Students

BY W. PAGET HIGGS, M.A., D.Sc.

FOURTH EDITION

CONTENTS

Symbols and the signs of operation. The equation and the unknown quantity. Positive and negative quantities. Multiplication, involution, exponents, negative exponents, roots, and the use of exponents as logarithms. Logarithms. Tables of logarithms and proportional parts. Transportation of systems of logarithms. Common uses of common logarithms. Compound multiplication and the binomial theorem. Division, fractions and ratio. Rules for division. Rules for fractions. Continued proportion, the series and the summation of the series. Examples. Geometrical means. Limit of series. Equations. Appendix. Index. 104 pages, 12mo, cloth, 60c.

See also **Algebraic Signs,** Spons' Dictionary of Engineering, No. 2. 40 cts.

See also **Calculus,** Supplement to Spons' Dictionary, No. 5. 75 cts.

Manual of Instruction in
Hard Soldering

WITH AN APPENDIX ON THE
Repair of Bicycle Frames
Notes on Alloys and a Chapter on Soft Soldering

BY HARVEY ROWELL

The flame, lamp, charcoal, mats, blow-pipes, wash-bottle, binding wire, chemicals, borax, spelter, silver solder, gold solder, oxidation of metals, fluxes, anti-oxidisers, oxidation of cases, the cone, oxidising flame, reducing flame, heat transmission, conduction, capacity of metals, radiation, application, the work table, the joint, applying solder, applying heat, the use of the blow-pipe, joints, making a ferrule, to repair a spoon, to repair a watch case, hard soldering with a forge or hearth, hard soldering with tongs, preserving thin edges, silversmith's pickle, restoring color to gold, chromic acid, to mend steel springs, sweating metals together, retaining work in position, making joints, applying heat, preventing the loss of heat, effect of sulphur lead and zinc, to preserve precious stones, annealing and hardening, burnt iron, to hard solder after soft solder. Tables of—specific gravity, tenacity, fusibility, alloys.

66 pages, illustrated, cloth, 75 cents.

For Soldering Receipts, Cements and Lutes, Pastes, Glues and such like, *see* WORKSHOP RECEIPTS.

NEW
EDITION "DE LUXE"
ON HEAVY PLATE PAPER

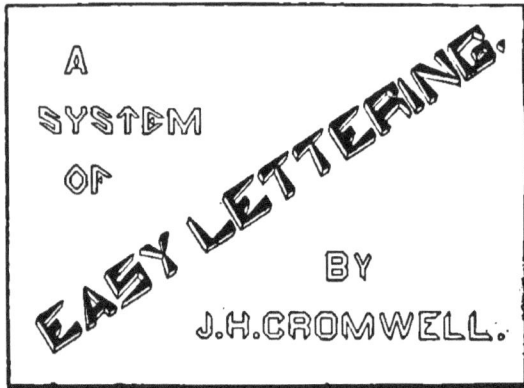

A
SYSTEM
OF
EASY LETTERING.
BY
J.H.CROMWELL.

ITS GOOD POINTS.

Very easy to learn.

A rapid method to become a good letterer with a little practice.

Very easy to lay out a line of words in STRICT PROPORTION, whether it be on a fence 500 yards long or on a drawing only a few inches across.

Good for draughtsmen who prefer neat lettering, yet something out of the ordinary.

It contains 26 pages of alphabets whose modifications are almost limitless.

One of the cheapest in the market.

This little book will be appreciated by draughtsmen who wish to use plain letters (and yet somewhat different from the ordinary run of letters) for the titles on drawings. The book will also be valuable and useful to any one who has had no practice in lettering, as the easy method given for forming the letters will enable a person to form the letters correctly, and with a little practice to do so quickly.— *American Machinist.*

Oblong, 8vo, cloth, 50 cents

USEFUL BOOKS

Barometer.—The barometrical determination of heights. A practical method of barometrical levelling and hypsometry, for surveyors and mountain climbers. By Dr. F. J. B. Cordeiro, U. S. N. 12mo, leather,   1.00

Dynamo.—Notes on the design of small dynamo, with complete set of drawings to scale. By G. Halliday. 79 pages, illustrated, 8vo, cloth, ... 1.00‡

Electric Bells.—A treatise on the construction of electric bells, indicators and similar apparatus. By F. C. Allsop. 131 pages, 177 illustrations, 12mo, cloth, ... 1.25

Electric Bells.—Practical electric bell fitting. A treatise on the fitting up and maintenance of electric bells and all their necessary apparatus. By F. C. Allsop. 170 pages, 186 illustrations, 12mo, cloth,   .   .   . 1.25

Electrical Notes.—Practical electrical notes and definitions, for the use of engineering students and practical men. By W. Perren Maycock, E.E. 286 pages, illustrated, 32mo, cloth, ... ·75

Electricity.—Comparisons between the different systems of distributing electricity. By Prof. Henry Robinson. 8vo, paper, .   . .80

Galvanometer.—A series of lectures on the galvanometer and its uses, delivered by Prof. E. L. Nicols, and used by him in his class at Cornell University. 112 pages, 76 illustrations, 8vo, paper, ... 1.00

Induction Coils, and coil making. A treatise on the construction and working of shock, medical, and spark coils. By F. C. Allsop. 172 pages, 124 illustrations, 12mo, cloth, 1.25

Measurements.—A systematic treatise on electrical measurements. By H. C. Parker. 120 pages, 96 illustrations, 8vo, cloth, . 1.00‡

Mining.—Electricity in mining. Its practical applications to all the different operations of drilling, blasting, lighting, haulage, pumping as applied to mining. By Prof. Silvanus P. Thompson, D.S.C. 45 pages, 11 illustrations, 8vo, paper, ...80

Phonograph.—The phonograph and how to construct it, and a chapter on sound, with full set of working drawings. 12mo, cloth, 2.00

Transformer.—History of the transformer, translated from the German. By F. Uppenborn. 60 pages, 31 illustrations, 12mo, .75

Transformer.—Transformer design, a treatise on their design, construction and use. By H. Adams. In the work the author has avoided as much as possible all historical matter and unnecessary mathematical problems, and has confined himself to practical experience. The work contains much information that will prove of value to the draughtsman, designer and electrical student. Second edition. 75 pages, 34 illustrations, 12mo, cloth, ... 1.50‡

SMALL ACCUMULATORS

How Made and Used

A Practical Handbook for Students and Young Electricians

EDITED BY PERCIVAL MARSHALL, A.I.M.E.

Contents of Chapters

80 pages, 40 illustrations, 12mo, cloth, 50c.

THE MAGNETO-TELEPHONE

ITS CONSTRUCTION,

Fitting Up and Adaptability to Every-Day Use

BY NORMAN HUGHES

CONTENTS OF CHAPTERS

80 pages, 23 illustrations, 12mo, cloth, $1.00. In paper, 50c.

EVERYBODY'S BOOK ON ELECTRICITY

PRACTICAL ELECTRICS

A UNIVERSAL HANDY-BOOK

ON

EVERYDAY ELECTRICAL MATTERS

FIFTH EDITION

CONTENTS:

Alarms.—Doors and Windows; Cisterns; Low Water in Boilers; Time Signals; Clocks. *Batteries.*—Making; Cells; Bichromate; Bunsen; Callan's; Copper-oxide; Cruikshank's; Daniel's; Granule carbon; Groves; Insulite; Leclanché; Lime Chromate; Silver Chloride; Smee; Thermo-electric. *Bells.*—Annunciator System; Double System; and Telephone; Making; Magnet for; Bobbins or Coils; Trembling; Single Stroke; Continuous Ringing. *Connections. Carbons. Coils.* —Induction; Primary; Secondary; Contact-breakers; Resistance. *Intensity* Coils.—Reel; Primary; Secondary; Core; Contact-breaker; Condenser; Pedestal; Commutator; Connections. *Dynamo-electric Machines.*—Field-Magnets; Polepieces; Field-magnet Coils; Armature Cores and Coils; Commutator Collectors and Brushes; Relation of size to efficiency; Methods of exciting Field-Magnets; Magneto-Dynamos; Separately excited Dynamos; Shunt Dynamos; Field-Magnets; Armatures; Collectors; Brush Dynamo; Alternate Currents. *Fire Risks.*—Wires; Lamps; Danger to persons. *Measuring.*—Non-Registering Instruments; Registering Instruments. *Microphones. Motors. Phonographs. Photophones. Storage. Telephones.*—Forms; Circuits and Calls; Transmitter and Switch; Switch for Simplex; etc., etc.

135 PAGES.   126 ILLUSTRATIONS.   8VO.

Cloth, 75 cents

RUHMKORFF
INDUCTION COILS

Their Construction, Operation and Applications, with
Chapters on

BATTERIES, TESLA COILS AND
ROENTGEN RADIOGRAPHY . . .

By H. S. NORRIE

● ● ●

CONTENTS OF CHAPTERS:

183 pages, 12mo, cloth, 50 cents.

THE VOLTAIC ACCUMULATOR

AN ELEMENTARY TREATISE
BY ÉMILE REYNIER
TRANSLATED FROM THE FRENCH
BY J. A. BERLY, C.E., A.I.C.E.

CONTENTS

202 pages, 62 illustrations, 8vo, cloth, $3.00.

CORRESPONDENCE INSTRUCTION . .

IN

ENGINEERING.

Steam Engineering,

Electrical Engineering,

Civil Engineering,

Mechanical Engineering,

MECHANICAL AND ARCHITECTURAL DRAWING,

Plumbing, Heating, Ventilation,

Chemistry, Metal Work, Mining.

THE

International Correspondence Schools,

SCRANTON, PA.

www.ingramcontent.com/pod-product-compliance
Lightning Source LLC
Chambersburg PA
CBHW021824190326
41518CB00007B/739